Discerning Palates of the Past

An Ethnoarchaeological Study of Crop Cultivation and Plant Usage in India

by

Seetha Narahari Reddy

INTERNATIONAL MONOGRAPHS IN PREHISTORY

Ethnorchaeological Series 5

Printed in the United States of America

ISBN 1-879621-36-3 (Paperback)
ISBN 1-879621-37-1 (Library Binding)

Library of Congress Cataloging-in-Publication Data

Reddy, Seetha Narahari, 1963-
Discerning palates of the past : an ethnoarchaeological study of crop cultivation and plant usage in India / by Seetha Narahari Reddy.
p. cm. — (Ethnoarchaeological series ; 5)
Includes bibliographical references.
ISBN 1-879621-36-3 (pbk. : alk. paper) — ISBN 1-879621-37-1 (lib. bdg. : alk. paper)
1. Indus civilization. 2. Agriculture—India—History—To 1500. 3. Millets—India. 4. Ethnoarchaeology—India. I. Title. II. Series.
DS425 .R39 2003
934'.01—dc21

2002154200
CIP

This book is printed on acid-free paper. ∞

International Monographs in Prehistory
P.O. Box 1266
Ann Arbor, Michigan 48106-1266
U.S.A.

FOR

MY PARENTS, Raja Reddy and Jaya Devi

and

MY HUSBAND, Brian Byrd

Table of Contents

List of Figures

(Continued on next page)

(List of Figures continued)

List of Tables

Acknowledgments

This manuscript, originally my doctoral dissertation, has gone through many versions before its current form. Numerous individuals have assisted me in this endeavor. The most ardent and consistent of my supporters through the years has been my husband, Dr. Brian Byrd, whose valuable criticism and suggestions have made an immense difference to this research. His unwavering support, encouragement, help, patience and love were ever present through all the phases of the work. I am ever grateful to him for standing by me every step of the way.

My doctoral studies at the University of Wisconsin, Madison, were supervised by Dr. Jonathan Mark Kenoyer. In an environment where homesickness, culture shock and loneliness were abundant, he made my years at the University of Wisconsin more tolerable through his generous support, both financial and emotional. His guidance and assistance in my research were equally generous and invaluable. Dr. Kenoyer's continued support and encouragement are deeply appreciated, as are his constructive criticism and suggestions on earlier versions of this document.

Special thanks to Dr. Henry Bunn, Dr. Joseph Elder, Dr. Gary Feinman, Dr. Christine Hastorf and Dr. Anatoly Khazanov for their comments, input and support of my research. My thanks to Dr. Margaret Schoeninger for her input and guidance in my carbon isotope studies. I am grateful to Dr. Gregory Possehl and Dr. Steven Weber for inviting me to work with them on their projects in Gujarat at Oriyo Timbo and Babar Kot. Over the years, my research has benefited from the comments, advice and assistance of the following individuals: Dr. William Belcher, Dr. Kuldeep Bhan, Mr. John Cook, Dr. Gary Feinman, Dr. David Harris, Frank Herman, Dr. Gordon Hillman, Ms. Charul Joshi, Dr. M.D. Kajale, Dr. Carol Kramer, Dr. Anatoly Khazanov, Dr. Thomas Levy, Dr. Richard Meadow, Dr. Heather Miller, Dr. V. N. Misra, Dr. Kathleen Morrison, Dr. Rafique Mughal, Dr. Urmila Pingle, Mr. Appa Rao, Dr. Marguerite Robinson, Kathleen Ryan, Dr. Carla Sinopoli, Dr. Glenn Stone, Dr. Kenneth Thomas, Dr. Massimo Vidale, Dr. Vishnu-Mittre and Dr. Wilma Wetterstrom. I would like to take the opportunity to thank and acknowledge Dr. Bob Whallon, who encouraged and supported the publication of this research.

My field research in India would not have been possible without the help and assistance of several individuals and institutions. I would like to thank the Archaeological Survey of India for permitting me to work at Oriyo Timbo and Babar Kot. Thanks to Dr. M. H. Raval and the Gujarat State Department of Archaeology. Special thanks to Dr. Y. M. Chitalwala of the Department of Archaeology, Western Circle, Rajkot, for his generous help and advice during my research in Gujarat. My ethnographic research in Gujarat greatly benefited by the generous help and expertise of farmer Naru Bhai and his wife of Babar Kot whose farm I studied. Many thanks to the late Mr. Raghav Reddy and the people of Rapanpalli and Sironcha in Andhra Pradesh for making the study of opportunistic cultivation possible. My thanks to Dr. Kuldeep Bhan for facilitating and assisting in the preliminary study of pastoral groups in Gujarat.

Funding for the ethnographic field research in Andhra Pradesh and carbon isotope analysis came from National Science Foundation Dissertation Improvement Grant (BNS 9024580). Funding for the ethnographic field research and analysis in Gujarat came from an American Institute of Indian Studies Junior Fellowship granted to me. In this connection I thank Dr. Joseph Elder, Dr. P. R Mehinderatta, and the staff at the Delhi office for their help and assistance during my fellowship year.

I thank my siblings and parents-in-laws for their encouragement, and my daughter, Asha, who has been a source of much inspiration to me. Last but not the least, I am deeply grateful to my parents for their unwavering support, encouragement and help. Special thanks are warranted to them for their input and help during my ethnographic research. I am especially grateful to both my parents for instilling in me the desire to pursue knowledge, regardless of time, culture or place.

1. Plants, Past and Present: An Introduction

The Harappan Phase of the Indus Valley Tradition of Pakistan and northwest India represents the earliest urban settlements in South Asia (Meadow and Kenoyer 1997; Kenoyer 1997b; Possehl 1997). This book explores the agricultural and pastoral subsistence infrastructure that characterized Harappan Phase social organization in Gujarat, northwest India, which is the southernmost region of the Indus Valley Tradition. The third millennium B.C., Mature Harappan Phase in Gujarat is characterized by urban centralization, craft specialization, and a diversified subsistence economy. The subsequent Late Harappan Phase of Gujarat is considered to have retained specific Mature Harappan attributes in a context of deurbanization, the development of new regional traditions, expansion of settlements into new areas, and hypothesized changes in subsistence regimes such as increased pastoralism. Elucidating the nature and causes of these developments in settlement patterns and subsistence strategies in the Gujarat Harappan has been the focus of considerable investigation (Bhan 1989, 1992; Patel 1997; Possehl 1986, 1992; Rissman 1985). Yet, until recently the precise nature of agricultural practices during this time period and the implications for diachronic changes in settlement patterns has remained largely conjecture (Weber 1991, 1998; Reddy 1994, 1997).

This book rigorously explores the economic role and importance of millet crops during the Harappan Phase in Gujarat. I hypothesize that millet cultivation may have supplemented existing nonfarming subsistence practices such as pastoralism, and played a crucial role in the context of changing subsistence economies and socio-political systems. Specifically, I was interested in whether millet crops were cultivated by occupants of particular sites or traded and brought into the settlement, and whether millets (regardless of whether they were imported or cultivated) were used as human food, animal fodder, or a combination of both. Three lines of evidence were employed to address this topic: ethnographic studies of crop processing, paleoethnobotany, and carbon isotope analysis. Initially, an ethnoarchaeological project was designed and implemented to model crop processing for archaeological application. This entailed developing an innovative methodology to isolate variables which distinguish different stages of crop processing in order to identify local crop cultivation in the archaeological record, and whether millets are processed differently depending on whether they are grown only as human food, both as human food and animal fodder, or when grown only as fodder. Next, paleoethnobotanical samples from varied contexts were recovered from two very different Gujarat Harappan sites. Then, the ethnographic crop processing model was rigorously tested through a detailed study of contextual associations within the archaeobotanical assemblages. Finally, these results were integrated with research on the role of wild and domestic fodder in pastoralist economies (including carbon isotope studies) to present a robust interpretation of plant cultivation, usage, and overall subsistence orientation during the Harappan Phase in Gujarat.

Chapter 2 sets the stages for the research questions and problems, and includes an introduction to the landscape of the study area with sections on the modern geography and protohistoric environmental setting. It presents the key issues and themes in the study of the Harappan Phase in Gujarat, northwest India, and their implications for the directions of new archaeological research and analysis. The context for Mature and Late Harappan subsistence settlement organization, particularly the strategies at play, namely those of agriculture and pastoralism are introduced. The concept of integration of agriculture and pastoralism is considered, suggesting that elucidating the degree of integration between economies is of critical importance in determining the subsistence system at play during the Mature and Late Harappan. This chapter also introduces the theoretical and methodological approaches that have been adopted.

The remainder of the book is dedicated to addressing the issue of defining and modeling plant usage through ethnographic studies and modeling, archaeological application and finally, the overall Harappan settlement subsistence systems in Gujarat are modeled and discussed.

The third chapter presents the ethnographic crop processing studies conducted in India, and includes a methodological account of the fieldwork. The objective of the ethnographic studies, the socioeconomic contexts of the study areas, and the sampling strategies are discussed. It also summarizes the results of the ethnographic crop processing studies and offers preliminary interpretations, particularly as they relate to archaeological interpretation of assemblages related to specific processing methods and stages. Chapter 4 presents the data analysis of the ethnographic materials, and offers models for archaeological application. This chapter examines the compositions of various crop byproducts and products at different processing stages, and models are constructed for the main crops and crop types studied, specifically tailored for archaeological application and testing. Chapter 5 focuses on the archaeobotanical research at Oriyo Timbo and Babar Kot in Gujarat, and includes a discussion of the methods used for collection, processing, and analysis of the archaeobotanical data. The archaeological application of the ethnographic crop processing models and interpretation of the archaeobotanical assemblages are presented in Chapter 6. Plant usage models and subsistence modeling for the two sites studied in this investigation are addressed. Chapter 7 discusses the modeling of animal diet and fodder acquisition, and includes the methods and results of carbon isotope analysis. An animal feeding model for the Harappan Phase in Gujarat is also presented which addresses seasonality, wild fodders versus cultivated fodders, and herd sizes. The final chapter of the book traces the broader implications of the results for Harappan subsistence and settlement in Gujarat, northwest India. The book concludes with a summary of the results and their implications for paleoethnobotanical interpretation, for the study of South Asian prehistoric agriculture, and to archaeology in general.

2. Archaeological Context and the Scope of Inquiry

Chapter 2 has two main issues of discussion: the archaeological context and the scope of inquiry. It initially presents the geography, environmental setting, and prehistoric cultural context for the research addressed in this book. Then, the prevailing interpretations and major research topics for the Harappan Phase are discussed. A brief summary of the Indus tradition is followed by a discussion of the Harappan Phase in Gujarat. Subsequently, the objectives of this study in the context of the Harappan Phase in Gujarat are presented, along with the scope of the problem and an integrated approach.

Geography and Environmental Setting

This study is set in northwest India, within the modern state of Gujarat (Figure 2-1). This is the westernmost state of India, and it has an extensive Arabian Sea coastline, and represents an intrusion of the Indo-Gangetic environmental conditions into the peninsula of India (Patel 1977). A distinctive feature of Gujarat is the strong seasonality inherent in its ecology. The modern state of Gujarat is comprised of two major geological zones, the Deccan Trap and alluvial riverine deposits. The Deccan Trap is a geological formation resulting from Mesozoic lava flows, which are some of the largest flows in the world (Patel 1977). Associated with this Deccan Trap are the black cotton soils, which are formed in situ and are extremely important for subsistence agriculture. A particular character of these soils which makes them excellent for dry cropping is their ability to swell and absorb substantial amounts of moisture. The Deccan Trap formation dominates the lithology of peninsular Gujarat (Saurashtra), south Gujarat and Kutch. In contrast, the alluvial deposits are more limited and mostly Pleistocene sediments deposited by the four main rivers of the region, the Sabarmati, Mahi, Tapi, and Narmada rivers. The active flood plains, however, are recent (post 10,000 B.P.) Holocene sediments.

The agricultural subsistence potential of the geography of Gujarat is synthesized by Patel (1977) who discusses the geography of the four main regions in Gujarat (South Gujarat, North Gujarat, Kutch, and Saurashtra) (see Figure 2-1). Gujarat is divided into these four regions based on geographical criteria and ethnic composition. South Gujarat is characterized by saline soils, and large tracts of pasture and grasslands in the west. High quality soils well suited for agriculture characterize the central areas (Wadia 1966), while North Gujarat is sandier and a much more arid plain. Relict Pleistocene sand dunes are present, and most of the rivers in this region are fed by the monsoon and ill suited for irrigation (Leshnik 1968). The third region, Kutch, is very arid and characterized by relict estuaries that are now salt flats created by the clay and sand deposits from minor rivers in the region (Gupta 1977).

Today, Saurashtra is the peninsula-like region of Gujarat with a central plateau and clayey black cotton soils known for their moisture retaining quality. However, during the Harappan Phase it is believed to have been insular. This study, including the ethnoarchaeological research and the archaeological sites of Babar Kot and Oriyo Timbo which are examined, is located in Saurashtra bound by the Gulf of Kutch on the north, the Gulf of Khambhat on the south, the Arabian Sea on the west, and the marshes of the Nal Depression on the east (Figure 2-1). The Nal Depression is of particular interest with respect to Harappan population movements. Spate and Learmouth (1967) have suggested that the Nal Depression (which is described as a land bridge connecting Saurashtra to the mainland of Gujarat) may have been flooded yearly until 1813, effectively creating an island of Saurashtra. Rissman (1985:57) refers to the Imperial Gazetteer of the Government of India (1908:170) in which 16th to18th century travelers' accounts allude to an Indus River branch flowing by the town of Khambhat, thus separating Saurashtra from the mainland. If this situation did indeed exist during the Harappan Phase, it would have had profound implications for the movement of populations from the Indus River valley in modern Pakistan to the Gujarat region, since it would have restricted movement during certain seasons except by boat.

Gujarat has a highly variable tropical climate, with most of North Gujarat and the northern fringes of Saurashtra having an arid climate (Patel 1977). This region is comprised of large grasslands and isolated patches of tropical thorn forest. South Gujarat has a sub-humid climate with considerable vegetation cover and tropical dry forests. The remainder of the Gujarat area has a semi-arid climate with dry deciduous forested areas and generally poor vegetation cover (Patel 1977). Seasonality is pronounced in Gujarat and includes three distinct seasons: winter, summer, and monsoon. The winter season extends from November to February, with December and January being the coldest months. The summer months of April and May are the hottest with maximum temperatures varying from 37.8o C in the south to 44.4o C in the north. The summer season ends with the onset of monsoonal rains in late June, and rains continue into September, with July and August being the wettest months of the year. The average precipitation per year is 60 cm; however, there is considerable variation within Gujarat. North Gujarat has an annual rainfall of 40 cm in the west and 80 cm in the east. Rainfall in South Gujarat varies from 80 to 100 cm annually, while Kutch has less than 40 cm annually. There is a great deal of variation in the rainfall of Saurashtra, ranging between 40 and 75 cm annually. Thus, there are distinct differences in vegetation from one region of Gujarat to another dependent on annual precipitation, temperature, and soils.

There is considerable homogeneity in the distribution of animals in Gujarat (Rissman 1985), although the densities may vary from one region to another. The Gujarat wild fauna includes *Antelope cervicapra* (blackbuck), *Axis porcinus* (hog deer), *Boselaphus tragocamelus* (nilgai), *Canis aureus* (jackal), *Canis lupus pallipes* (wolf), *Cervus axis* (spotted deer), *Cervus duvanceli* (swamp-deer), *Cervus unicolor* (sambar deer), *Equus hemionus* (half-ass), *Felis tigris* (tiger), *Gazella dorcas* (chinkara), *Hyaena hyaena* (common hyena), *Panthera leo* (Asiatic lion), *Sus scrofa* (wild boar), *Vulpes vulpes* (fox), and a whole range of birds and small mammals (Rissman 1985). A variety of domesticated animals (cattle, sheep, goats, buffalo, dogs, and chicken) are also present today. Cattle, sheep, and goats are used by both pastoralists and settled agriculturalists that inhabit the region.

In South Asia, variation in sedentary peasant agriculture is fundamentally distinguished based on the use of dry or wet farming, and whether a crop is an autumn (kharif) or spring (rabi) harvest. Dry farming is defined here as agriculture that is based on the natural rainfall, and it is characterized by the absence of irrigation. In contrast, wet crops are grown with the use of irrigation. Kharif or summer crops are monsoon crops, sown soon after the onset of the rains (June-July) and harvested in autumn. These include rice, millets, *Sesamum*, cotton, and jute. In contrast, rabi or winter crops are sown after the rains end and are harvested in the spring. These include wheat, barley, gram, mustard, and rape. A farmer can thus have two major crops on the same field each year.

Unlike wheat and barley, which are generally irrigated crops in South Asia, the various millets (summer crops) including Sorghum (*Sorghum bicolor*), finger millet (*Eleusine coracana*), bulrush millet (*Pennisetum typhoides*), proso or common millet (*Panicum miliaceum*), and foxtail millet (*Setaria italica*) are primarily dry farmed throughout the semi-arid tropics. Millets, often referred to as summer crops, tolerate a wide range of soil types and have a relative short warm-weather growing season. Today they are planted in South Asia principally on the Deccan plateau and in Gujarat. Gujarat's agriculture is primarily dry farming (85%) as a result of the rich black cotton soils and sufficient annual rainfall. Millets are also cultivated in Baluchistan, Rajasthan, and along the western margins of the Indus Valley using moisture from summer floods (Patel 1977). Agricultural practices in Saurashtra are similar to the rest of Gujarat, with dry farming being the prominent subsistence economy, supplemented by "pastoral nomadism" by different sects of the population.

Indus Valley Tradition and the Gujarat Harappan

Since its discovery in the 1920s, the Indus Valley Civilization has been the focus of considerable research and debate, particularly since it is South Asian's earliest urban society (Chakrabarti 1984; Jacobson 1987; Kenoyer 1991). The discussions of early urbanization have been based on extensive excavations at both large and small sites, and regional surveys (Bisht 1990; Dales and Kenoyer 1990; Jarrige 1986; Kenoyer 1989, 1991, 1997a, 1997b; Possehl 1980; Possehl and Raval

1989; Rissman and Chitalwala 1990). Investigations in the Cholistan Desert by Mughal (1982, 1988; 1990), in Punjab (Pakistan) by Dales and Kenoyer (1990, 1991) and Meadow and Kenoyer (1997), in Southern Indus by Dales (1979) and Fairservis (1973), in Baluchistan by Jarrige (1988, 1997), and in Saurashtra by Bhan (1989, 1992), Possehl (1980, 1992), Possehl and Raval (1989), Rissman and Chitalwala (1990), and Weber (1990), have all contributed to deepening our understanding of the Indus Valley cultural tradition.

South Asian archaeology (especially the Indus Valley Civilization) has been subject to a great deal of criticism from scholars of other regions of the world, especially with respect to interpretative theory. Much of the criticism points out the lack of extensive regional survey data from South Asia, and the tendency of many South Asian scholars to explain prehistoric diversity in South Asia in terms of similarities and differences with other regions of the world, such as Mesopotamia, Egypt, and Europe (Gupta 1979; Lamberg-Karlovsky 1982). Until recently progressive evolutionary theories have been common in the interpretations of South Asian prehistory, where specific culture change is viewed as one event in an ordered sequence of events. However, the limitations of such approaches are now being recognized and other explanatory models are being offered (e.g., Belcher 1997; Jarrige and Meadow 1980; Kenoyer 1991, 1992, 1997a, 1997b; Meadow 1996; Mughal 1982; Possehl 1980, 1992; Reddy 1994, 1997; Shaffer 1991; Weber 1989, 1997).

Two key controversial issues in the archaeology of the Indus Valley Civilization are chronology and terminology. The terminology that is most widespread in the archaeological literature for the period between 4000 B.C. and 1500 B.C. in South Asia includes the phases Pre-Harappan, Early Harappan, Mature Harappan, Late Harappan and Post Harappan (Kenoyer 1991). Similarly, terms such as Early Indus and Mature Indus are also used but without precise definitions. In addition, all the terms are used in reference to the urban character of the civilization at its zenith, and some scholars have recognized the limitations of such terminology (Kenoyer 1991; Shaffer 1991).

Shaffer (1991) proposed an explicit alternative, interpreting the Indus Valley archaeological record as a cultural tradition. Central to his comprehen-

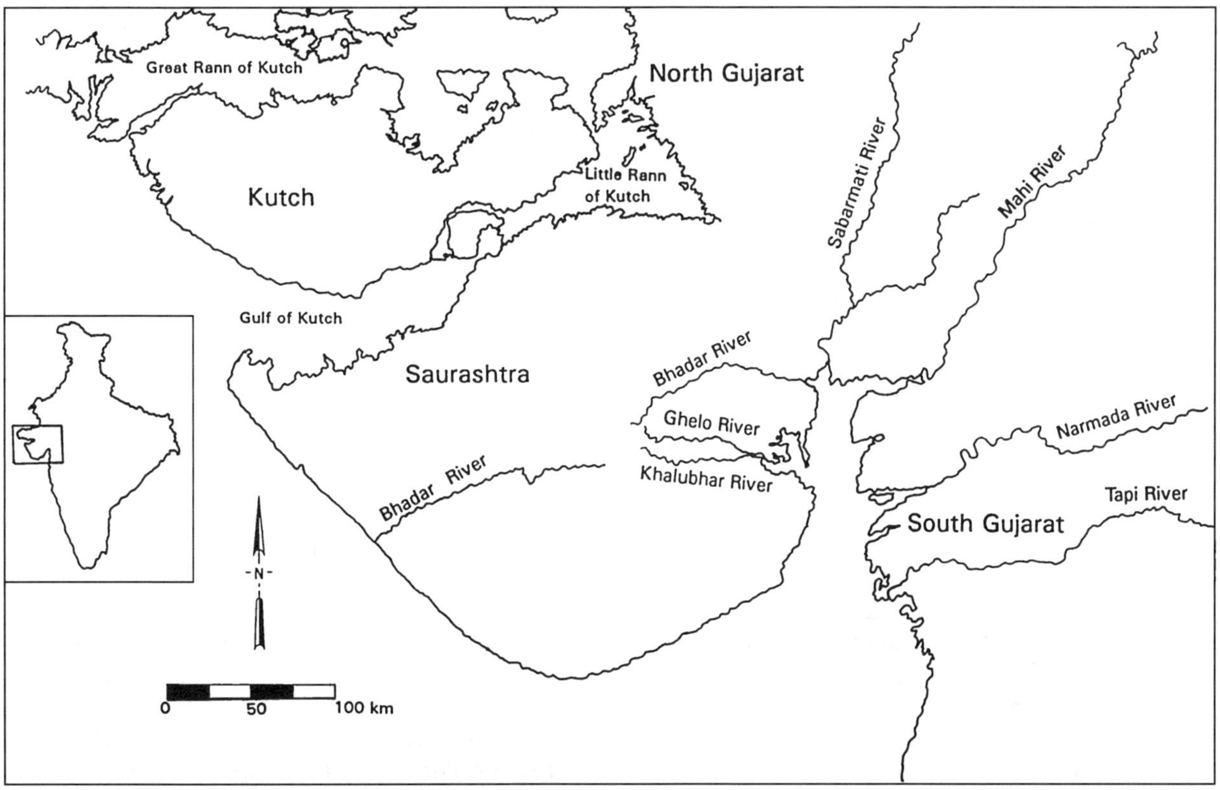

Fig. 2-1. Map of Gujarat showing the geographical regions.

sive terminology is the heuristic concept of 'tradition,' which he defines as "persistent configurations of basic technologies and cultural systems within the context of temporal and geographical continuity" (Shaffer 1991:442). In this conceptual framework, Shaffer (1991) identifies three major cultural traditions in Northwest South Asia. Each tradition is divided into temporal eras (which are unrelated to geological eras) and phases. There are four distinct eras for the Indus Valley Tradition: Early Food Producing Era; Regionalization Era; Integration Era; Localization Era. Based on ceramic data, Shaffer identifies phases within each era. His framework is an effective approach to the massive archaeological record of the Indus Valley Civilization. Of concern to this project are the Integration and Localization Eras in the Indus Tradition. The Harappan Phase of the Integration Era begins between 2700 B.C. and 2500 B.C. and occurs in the core regions of the Indus Valley and Gujarat (Kenoyer 1991). The date for the end of this Harappan Phase is debated, and it is possible that it could continue up to 1900 B.C.

Kenoyer (1991) provides an excellent comprehensive discussion of the Integration and Localization Eras of the Indus Valley Tradition. For the purpose of this study, which is focused on the Gujarat Harappan, the terms "Mature Harappan" and "Late Harappan" are considered the most appropriate since most scholars working in Gujarat have continued to use this terminology, and acceptance of Shaffer's (1991) framework is not universal. The term 'Mature Harappan' in Gujarat is equivalent to Shaffer's Harappan Phase (2500 - 2000 B.C.) in the Integration Era of the Indus Tradition, while the term 'Late Harappan' in Gujarat can be broadly equated to the Rangpur Phase (2100 - 1300 B.C.) of the Localization Era in the Indus Tradition (Shaffer 1991). It should be noted that there are other chronologies proposed for the Indus Tradition, and they vary slightly from Shaffer's (1991) chronology (e.g., Kenoyer 1991).

Thus, the Indus Valley Tradition as defined by Shaffer (1991) and applied by Kenoyer (1991) includes cultural development in the Greater Indus Region from around 6500 B.C. until 1500 B.C. or later. This time span starts with food production, and includes regionalization, integration and finally localization. The focus of this book is on the Integration and Localization Eras, particularly of the Harappan culture in Gujarat, India. The urban state-level society of the Harappans was diverse and complex, with distinct regional expressions. It is important to reiterate that the different regions of the Harappan sphere of influence (Gujarat, Punjab, Sindh, and the main Indus River Valley) were not in isolation from each other (Kenoyer 1991; Meadow 1989; Possehl 1980). For example, Possehl (1979) proposed that pastoral nomads or other highly mobile (itinerant) occupational specialists moved between the different regions of the Indus civilization during the Mature Harappan Phase. Therefore, in spite of spatial distinction, there is considerable homogeneity in the Harappan culture communities from these various regions. There are also distinct stylistic and cultural signatures or markers that distinguish specific regions.

With respect to the subsistence orientation of the Indus Valley Tradition, Meadow (1989, 1996) has argued for two distinct agricultural systems for the northwestern region of South Asia during the prehistoric and the protohistoric periods. The first involved the establishment between 7500 - 5500 B.C. of a farming complex based principally on the rabi (winter sown, spring harvested) crops of wheat and barley. The second, 2500 - 1500 B.C., entailed the addition of kharif (summer sown, fall harvested) crops such as Sorghum and other millets (Hutchinson 1977; Meadow 1989). Possehl (1986) argued that the development of summer, rain fed, 'millet'-based agriculture in South Asia took place largely on the eastern and southern margins of the Indus Valley during the Harappan Phase. These areas have sufficient monsoon rainfall to support the dry farming of hardy, warmth-loving cereals. As Possehl (1986) has noted, the development and adoption of agricultural practices that used monsoon rainfall, opened up vast areas of Gujarat and the Deccan for dry farming.

Protohistoric Environmental Setting

The protohistoric climate and environment are variables that played important roles in many theories regarding the Harappan Phase of the Indus Valley Tradition (e.g., Raikes and Dyson 1961; Singh 1971). Both are viewed as limiting factors in cultural adaptation and culture change in South Asia by many scholars working in this area. Therefore, interpretations of the climate and environment during the Harappan Phase of the Indus Valley Tradition have drawn considerable interest and controversy in South Asian archaeology literature. The earliest scholars, such as Marshall (1931), Dikshit (1938), Mackay (1943) and Piggott (1962), argued that the protohistoric climate and environment were distinct from today. Since then, a series of studies have offered conflicting views on the climate. The most notable disagreement concerns the amount of rainfall, and interpretations have ranged from greater rainfall, to lesser rainfall or rainfall the same as today.

Singh (1971, 1977) is the most prominent proponent of the earlier theories, which favored greater rainfall during the Mature Harappan of the Indus Valley Tradition. He used pollen sequence data from lakes in Rajasthan, India, to argue that the climate was moister in the past due to increased rainfall (as compared to today) in Northwest South Asia during the Harappan Phase, which then decreased during the early parts of the second millennium B.C. Singh's sequence has been cited in climatic/environmental deterministic theories for the decline of the Indus Valley Tradition in Northwest South Asia (Agrawal and Sood 1982; Singh 1977).

In contrast, there are a number of scholars who argue that the climate during the Harappan Phase was similar to present times (Fairservis 1956; Hegde 1977; Raikes and Dyson 1961; Meadow 1989; Possehl and Raval 1990; Possehl 1992; Rissman and Chitalwala 1990). The data on climate and environment that they use to support their interpretation also are derived from pollen studies. However, in general the climatic evidence for this area of the globe is very poor, primarily due to the lack of suitable deposits for long-term pollen cores. Kenoyer (1991) notes that even if there were changes in rainfall, the area is so vast that it would not have affected all regions uniformly. Thus, scholarly debate on climate change or stability during the Harappan times remains unresolved since no definitive evidence has been obtained. If there were differences in rainfall, then what effect did they have on the distributions of plant communities and the viability of different economic strategies? Shaffer and Liechtenstein (1989) contend that the northwest portion of South Asia did experience a series of climatic changes during the Holocene. However, the extent, timing, and duration of these fluctuations and cultural impact/responses are poorly understood.

For Gujarat, India, the available localized climatic data are generally interpreted as indicating no major differences between the third/second millennia B.C. and today (Meadow 1989; Rissman and Chitalwala 1990). Based on pollen data, it has been argued that the environment in Gujarat has changed little since the mid-Holocene. This position has been articulated in detail by Weber (1989) who refers to the works of Agrawal and Pande (1977), Mann (1955), Allchin and Goudie (1971), Allchin et al. (1978), Bhandari (1974, 1978), Bharadwai (1961), Gupta (1971a, 1971b), Gupta and Sharma (1982), Dodia (1988), Dodia et al. (1982), and Meher-Homji (1989). For the purpose of this study it is assumed that there have been no major fluctuations in Gujarat. The only difference would be the insular nature of Kutch, and possibly Saurashtra, which would have significant effects on the settlement of areas and movements of populations.

Gujarat Harappan: Definition and Broader Issues

Gujarat lies approximately 970 kilometers southeast of the site of Harappa in the Indus Valley Harappan Phase core area. The geographical region referred to as the modern state of Gujarat in India is presently the southernmost extension of the Harappan cultural tradition (Figure 2-2). The Harappan tradition of Gujarat is a distinct regional expression, and excavations at a number of sites have revealed a long history of Harappan presence in this region (Possehl and Raval 1989; Herman 1997a). Though on the periphery of the Harappan core area, Gujarat stands as an important region for understanding the Harappan cultural sphere of interaction (Possehl and Herman 1990). Possehl and Raval (1991) assert that Gujarat appears to have been settled by Harappans from the greater Indus region around 2400-2300 B.C., based on radiocarbon dates from the sites of Lothal and Surkotada (Figure 2-2). However, Possehl (1992) and Herman (1997a) note that three sites, Prabhas Patan (Somnath) (Sankalia 1972), Nagwada (Hegde et al. 1988) and Dholavira (Bisht 1990), offer scant but intriguing evidence of the presence of pre-urban populations in north Gujarat and Saurashtra. For example, Nagwada has ceramic assemblages indicative and reminiscent of Amri in Sindh (Pakistan), an Early Harappan site dating to 3000-2600 B.C., and ceramic indicators of the Urban phase. There are over 500 sites in Gujarat that have been identified as containing a component related to the Harappan tradition ca. 3000-1300 B.C. (Possehl 1992). The Kutch area of north Gujarat has been viewed as the corridor of cultural exchange and flow between the Indus Valley, specifically Sindh, and Gujarat. The site of Dholavira, situated on an island in the Rann, is currently interpreted as a "staging point" for cultural exchange and population movement during this period (Bisht 1990; Possehl 1992).

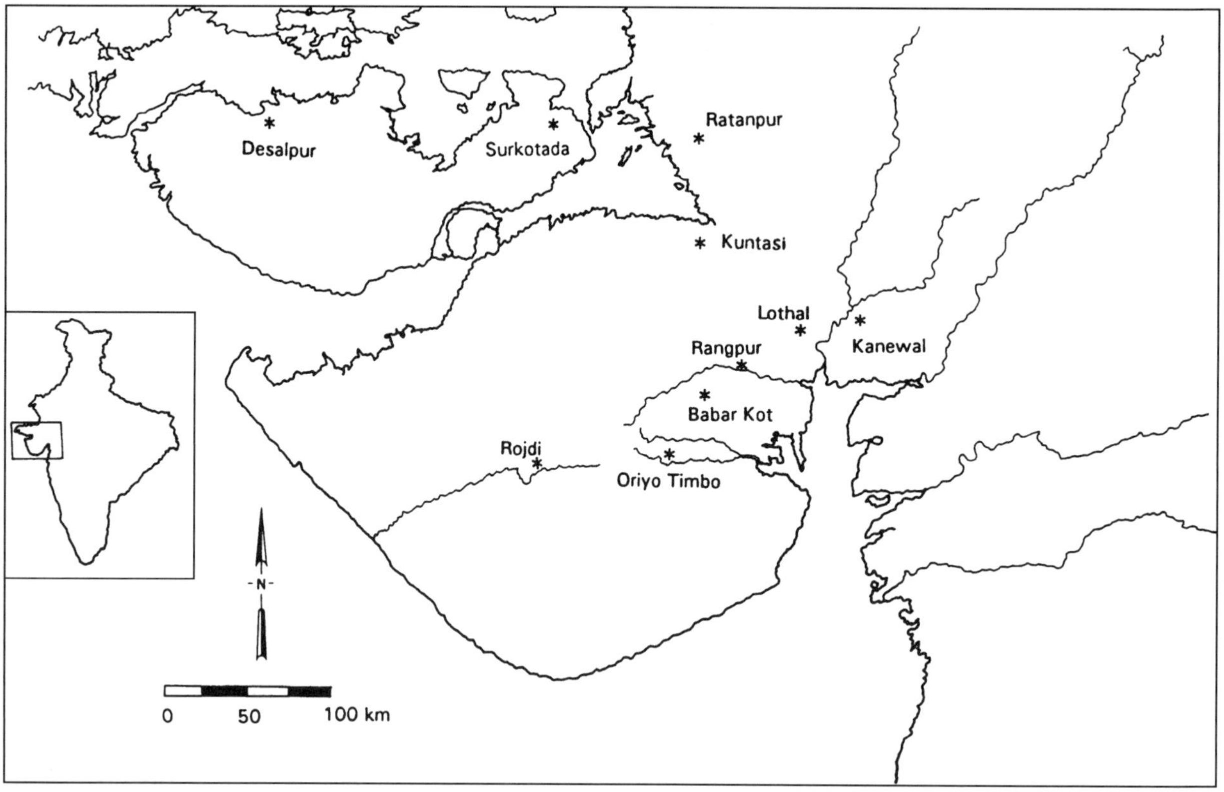

Fig. 2-2. Map of Gujarat showing the archaeological sites mentioned in the text.

At present, the Harappan Phase in Gujarat is burdened†with a variety of terminological frameworks used to define the urban and post-urban cultures of this period. For example, some of the terms used for the Mature Harappan include Urban Harappan and Sindhi Harappan, while the Late Harappan has also been called the Post Harappan, Post Urban Harappan, Sorath Harappan, Prabhas Harappan, and Gujarati Harappan. Intermingled with these terms are site-specific relative phase terms based primarily on ceramic chronology. The lack of a single, well-dated chronological sequence is a major difficulty inhibiting advances in the study of the Gujarat Harappan. One reason for this lack of resolution is that until very recently ceramic chronology was the fundamental and often the only form of dating used for the Gujarat Harappan. Therefore, the Gujarat Harappan literature is filled with discussions of ceramic chronological sequences, and typically the Mature Harappan site of Rangpur is the type site (Herman 1997a, 1997b). Other sites are then equated and compared to the Rangpur ceramic sequence, and the reader has to contend with chronological terms such as Lothal B, Rangpur IIB/C, Rojdi B, and so on. Radiocarbon dates obtained in the last decade or so have helped resolve some of the confusion, and the Rangpur sequence is generally equated to the major Harappan Phases (Table 2-1).

There are important limitations, however, in the use of the Rangpur ceramic chronology. Herman (1997a, 1997b) argues that it is a fallacy to continue relying on the Rangpur correlations since the Rangpur data has severe flaws particularly related to sampling errors, and this sequence has been uncritically applied to the Harappan in Gujarat. Much of the chronology of the Gujarat "Harappan" is based on a stylistic comparison of ceramics from various sites, and not complemented by ample radiocarbon dating. As a result, the literature is severely hindered by the classification of sites through relative dating (ceramics being the primary artifact category being used for stylistic dating).

Recent assertions by Possehl and Raval (1990), Possehl (1992) and Possehl and Herman (1990) that the Gujarat Mature Harappan has two expressions, Sorath and the Sindhi, have added on new cultural identities to address and define within the existing Gujarat Harappan terminology. The Sorath Harappan is considered a distinct local manifestation, while the Sindhi Harappan sites

Table 2-1. Gujarat Chronological sequence.

RANGPUR	PHASE	DATES
IIA	Mature Harappan	(?)2500-2000 B.C.
II B-C	Late Harappan or Initial Late Harappan	?2000- ?1700 B.C.
III	Post Harappan or Final Late Harappan	?1700 - ?1400 B.C.

are seen as being directly associated with the "core area" of the urban Indus system. Possehl (1992) concludes that all the Late Harappan sites in Saurashtra are Mature Harappan sites of the Sorath Harappan class. Of interest is the emerging complex picture of cultural affiliation of the peripheral Gujarat Harappans to the core Indus Valley (see Herman 1997a for full discussion).

The Sorath Harappan in Gujarat has been first recognized and redefined by Possehl and Raval (1989) and Possehl and Herman (1990) who noted that Rojdi and many other sites in Gujarat represent a newly discovered regional expression of the Mature Harappan Phase, which clearly is a part of the larger Harappan cultural entity centered in the Indus Valley. Previously excavated Mature Harappan sites in Gujarat such as Lothal, Rangpur, Surkotada, and Desalpur have a typical Mature Harappan material culture inventory that includes inscribed seals, distinctive ceramics, metal works, beads, and substantial public architecture. This aspect of the Indus civilization in Gujarat is referred to as the 'Sindhi' Harappan by Possehl and Raval (1989), since it closely resembles materials from the Indus core sites, which are found in Sindh, Baluchistan, Punjab, North West Frontier Province, and Haryana. In this project, the term 'core area' is used in place of Possehl and Raval's (1989) term 'Sindhi'. Radiocarbon dates place this phase between 2500 B.C. and 2300 B.C.

The Sorath Mature Harappan sites in Gujarat, particularly in Saurashtra, are very different from the core area Harappan sites in that their material culture inventory is relatively less diverse, there is the noted absence of stamp seals, ornaments are rare, architecture is not elaborate, and site size is small (about 5.3 hectares). Possehl and Raval (1989) argue that these sites represent a distinctive regional expression of the Mature Harappan Phase, which they termed 'Sorath' Harappan (Possehl and Herman 1990). They argue that it is stylistically divergent from the core area Mature Harappan (as known from the Mature Harappan sites in Sindh and Kutch) but still part of the Harappan integration sphere. Possehl (1992) states that the Sorath Harappans were farmers and herders, and their sites (with the exception of Nageswar) have very little evidence for craft activity or "industrial" production of metals, beads, bangles, or related items. The Sorath Harappan in Saurashtra is characterized by different types of sites; most are small (e.g., Zekhada and Valabhi) comprised of low mounds, simple scatters of pottery and ranging from 0.5 to 1.5 hectare in size (Possehl 1992:129). These have been interpreted as temporary encampments of migratory herding peoples by Possehl (1992) and others. The other distinct category includes the larger deeply stratified sites such as Rojdi, Prabhas Patan, and Babar Kot, with substantial fortification walls, and ranging in size from 2 to 7 hectares. These larger sites are interpreted as permanent villages, occupied year round with agriculture and animal husbandry being an important component of their economy. It appears that Possehl (1992) uses the terms Sorath and Sindhi Harappan as implying Urban Phase Harappan with specific cultural meaning. It is unclear whether this clarification continues into the Post-Urban phase, where a Sindhi and Sorath Late Harappan can also be delineated.

Thus, Possehl and Herman (1990) suggest that Harappan populations were extensive in Gujarat at an earlier date than had previously been expected (ca. 2500-2200 B.C.). Therefore, the original interpretation that during the Mature Harappan Phase small numbers of Harappans migrated to Saurashtra from Sindh (Possehl 1976) has been replaced by a more complex picture comprised of local ('Sorath') and non-local (core area Harappan) cultures. An indigenous line of development within Gujarat may have begun around 3000 B.C. (Possehl and Raval 1989), and the core area Harappans may have then come into Gujarat, particularly Saurashtra, as a distinctive and separate migration about 2400 B.C. In spite of these recent developments, the Harappan Phase in Gujarat is yet to be fully understood.

One important issue in South Asian archaeology that has remained unanswered deals with the causes for a distinct 'deurbanization' of the greater Indus Valley beginning about 2000 B.C. (Meadow 1989). Along with this phenomenon was the emergence of distinctly local manifestations of material culture that are particularly striking when compared to the extent of integration evident during the Mature Harappan Phase. There are also significant shifts in settlement patterns including the eventual abandonment of some very large sites (including Mohenjo-daro and Harappa), and an increase in smaller settlements throughout the region and particularly in the east Punjab (Joshi et al. 1984; Shaffer 1981) and Gujarat (Possehl 1979). Some abandonments and realignments have been attributed to changes in hydrological regimes in Punjab (Mughal 1982; Wilhelmy 1969) and in Sindh (Flam 1981). New data from Mohenjo-daro suggest that the process of change that brought about the eventual abandonment of the site began in the latter part of the third millennium B.C. (Possehl 1997) .

Mughal (1990) argues that the Late Harappan culture Jhukhar of lower Sindh had contacts with Lothal A-B and Rangpur IIA-IIB of Saurashtra, Gujarat during the Late Harappan. He suggests that near the onset of the Late Harappan the urban interaction sphere did not fully break down, instead contacts and trade continued. Subsequently, the two regions took different courses of cultural development. Alternatively, recently Possehl (1997) argued that settlement data from the ancient Sarasvati River, Gujarat, and northwestern India suggest that there was no general eclipse or decline of the Indus Civilization but that a process of deurbanization occurred with a shift in the general distribution of population to the east.

The task of elucidating the changes and continuity in Gujarat from Mature Harappan to Late Harappan is challenging, and scholars have offered different perspectives (e.g., Bhan 1989; Herman 1991, 1997a). Herman (1991, 1997a) argues that there is a distinction between Harappan settlements of Kutch and Saurashtra beginning during the Early Harappan. The Harappan economy in Kutch included extensive production and trade of valuables throughout the Harappan interaction sphere. The Harappan emphasis in Saurashtra, in contrast, was on herding and farming, and the interaction sphere did not extend much outside of Gujarat. Thus, regionality and locality were more important for the Harappans in Saurashtra, as compared to the Harappans in Kutch who were involved in trade outside the Gujarat area from the early Harappan times. According to Herman (1991), the subsistence strategies were similar, with both areas being reliant on cattle herding and millet subsistence. With the intensification of the urban process during the Mature Harappan Phase, factory settlements were set up in Saurashtra by the Harappans of Kutch. In the Late Harappan (starting around 2000 B.C.) localization became heightened and the local interaction between Kutch and Saurashtra intensified owing to the breakdown of interaction spheres with the core areas to the west and outside the Gujarat region. Herman (1991) emphasizes the Kutch/Saurasthra dichotomy, but does not elaborate on the relationship between settlement patterns and subsistence strategies, which is particularly important in a semi-arid/arid environment. Herman (1997a) maintains that Saurashtra and the North Gujarat Plains remained peripheral to the Harappan culture throughout the sequence. During the Late Harappan, Kutch continued to function as an urban system, and peripheral Saurashtra increasingly acquired more traces of urbanism. Herman (1997a:105) suggests that there might have been a power shift from Kutch to Saurashtra during this late stage. Of intrigue is the abandonment of Kutch during the Post-Harappan times (post 1800-1900 B.C.). Thus, it is becoming increasingly clear that the Gujarat Harappan developed and evolved independent of the core area, and further investigations are imperative to clearly elucidate regional changes.

Culture change at the end of the second millennium B.C. from the Mature to the Late Harappan was not uniform and even paced as previously believed. Given the dearth in radiocarbon dating of Late Harappan sites, the time-range covered by this cultural phenomenon is not clear, however, it is evident that the time scale and processes involving the localization of the Harappan urban phase were specific to different regions. In general, the Gujarat Mature Harappan is characterized by urban centralization, craft specialization and a diversified subsistence economy, while in contrast the Gujarat Late Harappan is defined by the retention of specific Mature Harappan attributes but in a context of de-urbanization, expansion of settlements into new regions, changes in subsistence regimes, and the development of new regional traditions in ceramics and other aspects of material culture.

Gujarat Harappan Settlement Patterns

Extensive survey data from Gujarat indicate that during the second millennium there were several significant changes in the nature and distribution of settlements of the Harappan integration sphere (Bhan 1989, 1992; Possehl 1980, 1992). This period witnessed, among other things, an increase in the number of small sites coupled with the possible or partial abandonment of the larger urban sites, a decrease in elite goods (such as seals, beads, and metal works), and the decreased use of the Harappan weight and writing systems. These changes have been interpreted as being due to the 'collapse' or decline of the Harappan state, and a decreasing complexity of political and social systems as compared to the earlier Mature Harappan.

Bhan (1989) examined variation in the settlement pattern between Kutch and Gujarat from Mature to Late Harappan in relationship to subsistence strategies. He suggests that Mature Harappan settlements in Gujarat are significantly infrequent compared to Late Harappan settlements, and most are situated in the Kutch region, and he also makes a distinction between the initial phase and the final phase of the Late Harappan. He uses an expansionist theory to argue that the Kutch Mature Harappan sites were administrative settlements, while those in Saurashtra (such as Nageswar and Lothal), were manufacturing satellite settlements. Thus, the Mature Harappan settlement pattern developed mainly to facilitate trade, access to raw materials, and administrative activities, and not solely due to subsistence related purposes. He also suggests that the establishment of satellite settlements in Saurashtra increased toward the end of Mature Harappan with a deterioration of the integration with the core area. Additionally, Bhan (1989) suggests that two different categories of settlements developed during the Late Harappan in Gujarat; abundant small villages or dry season pastoral camps with simple round huts, and fewer but larger sedentary farming settlements with permanent mud brick or stone rubble architecture. Although there are no artifactual ties between these types of settlements, the two-tier subsistence system has great interpretive value for the studies of Late Harappan settlement and subsistence systems in Gujarat.

Bhan (1989) and Rissman (1985) attributed the fluctuation in site count during the Late Harappan to changing mobility mainly due to pastoralism. Rissman (1985) suggested that there is a shift from cultivation to pastoral production in the final phase of Late Harappan. Both Bhan (1989) and Rissman (1985) argue that the need for the maintenance of large herds exerted pressure on mobility, and thus the location of settlements in traditional pasture lands was an inevitable outcome in the context of such an economy. This trend is supported by the concentration of the North Gujarat/Kutch Late Harappan settlement in the western part where the agricultural potential is low but excellent grasslands are optimal for a pastoral economy.

Subsequently, Bhan (1992) reformed his site categorization, and included the initial phase of Late Harappan into the Mature Harappan, based on Possehl and Raval (1989) and Possehl (1992). The resulting picture is rather unclear, and greater precision and substantive additional dating and investigations are needed, as cautioned by Bhan himself, to clarify the temporal distinctions within Late Harappan in Gujarat. Note however, that Possehl and Mehta (1994) abandon earlier perceptions of limited number of large Mature Harappan sites and high number of small Late Harappan sites; and instead now propose a high number of Mature Harappan sites and fewer Late Harappan sites. Herman (1997a, 1997b) argues that the precise number of sites in each period will remain unknown unless good chronological markers are found, and he cautions the use of the Rangpur ceramic sequence in assigning a chronological term to sites in the absence of absolute dating. Given this, the site counts by Possehl and Mehta (1994) and Bhan (1989, 1992) have serious limitations in reconstructing the second millennium landscapes of Gujarat. As made apparent by Dhavalikar (1993), this Sorath/Sindhi Harappan distinction and interpretation of the Harappan tradition in Gujarat is not widely accepted. The obvious limitations to this theme include lack of temporal resolution with the absence of adequate absolute dating of these sites. Thus, it is clear that the debate on settlement patterns during the Gujarat Harappan is constantly evolving.

Gujarat Harappan Plant Subsistence Strategies

A subsistence pattern based principally on the winter crops of wheat and barley, and on domestic cattle was established in the Greater Indus Valley by the sixth millennium B.C., and it formed the agricultural foundation for the Indus Civilization (Meadow 1989, 1996). The subsequent diversity of the Harappan subsistence systems is a much-debated topic. Subsistence studies of the third and second millennium have focused on the apparent introduction of millets and associated secondary 'revolution' in the subsistence systems. Meadow (1989) based his argument on faunal data since horse and donkey first appear at Harappan sites at the start of the second millennium. Meadow (1996:398) suggested that millets were included in this corpus of new domesticates. Since then, however, Sharma (1992) has confirmed the presence of domesticated horse during the Mature Harappan, and more specifically at the site of Surkotada in Gujarat (note however that there is much debate and a lack of consensus on the identification of this genus). In contrast, Weber (1998) suggests that African millets came into South Asia at two points in time, first in the third millennium B.C. and the second introduction in the second millennium B.C. Weber (1989, 1990), through his archaeobotanical research at the Harappan site of Rojdi in Gujarat, has demonstrated that millets were present in the area from the mid-third millennium. It is plausible that the introduction and cultivation of millets in this semi-arid region of northwest India aided in the proliferation of settlements since they are more arid adapted crops and grow in wider range of soils as compared to the winter crops.

There is meager evidence of winter crop cultivation in Gujarat in the form of a few barley grains from the lowest levels of Rojdi (Weber 1989). Weber (1989) argues that barley could have been traded instead of having been grown locally. Additionally, the evidence of agriculture from sites prior to about 2500 B.C. is non-existent, mainly because the importance of studying subsistence economy and retrieving archaeobotanical materials has only recently been recognized. As the range of questions being asked about the relationships between people and plants widens, archaeologists are rapidly moving beyond descriptive lists of species (Hastorf and Popper 1988). The focus has shifted to total assemblages and additional lines of evidence such as ethnographic studies, bone chemistry, soil geology, and phytolith studies are being investigated and integrated into interpretive models (Pearsall 1982; Reddy 1991a). With the exception of Rojdi, and the two sites studied for this project (Oriyo Timbo and Babar Kot) there are very few Gujarat Harappan sites from which any archaeological plant remains have been recovered and studied. Small quantities of plant materials were also recovered from the Gujarat Harappan sites of Surkotada, Lothal and Rangpur (Ghosh and Lal 1963; Vishnu-Mittre and Savitri 1982), but the sampling strategy and methods were not intensive enough to answer the major question regarding the Mature and Late Harappan in Gujarat.

To what extent the practice of agriculture itself may have been an indigenous development in Gujarat rather than introduced from elsewhere is impossible to say at this point in time. Not only is very little known about early agriculture in the Gujarat region, but there are no plant remains reported from any pre-agricultural (Mesolithic) sites in the area. The limited archaeobotanical samples available from Mature Harappan sites in Gujarat include 'millets'. Since millets such as *Panicum miliare, Setaria italica, Eleusine coracana, Sorghum bicolor,* and *Pennisetum typhoides* are considered to be non-native and brought into Gujarat from elsewhere, the timing and dynamics of this process are of considerable interest (Weber 1989, 1990, 1998). Other evidence of agriculture in the Gujarat Harappan comes from impressions on pottery, the presence of a granary at Lothal, and the nature of settlements. The size of the larger Harappan towns (and indirectly the population) in Gujarat made plough agriculture a necessity. However, no ploughs have been found from Harappan sites in Gujarat, although a possible ploughed field of the pre-Harappan levels of Kalibangan is often cited as evidence. Other evidence of agricultural practices includes a seal found at Lothal, which depicts an object interpreted by Rao (1973) as a seed drill; however, it is an unusual shape for a seed drill.

The data on Gujarat Harappan agriculture (both Mature and Late) prior to this research project comes from select few sites and thus placed constraints on interpretive modeling. Additionally, since there is no information on agricultural practices prior to 2500 B.C., the origins of subsistence strategies of the Harappan are further complicated. The general consensus regarding Gujarat Harappan subsistence is that animal herding was an important activity during the Mature Harappan and that this strategy increased in the Late Harappan. In addition, summer crop cultivation was practiced and was supplemented by wild plant and animal exploitation, and winter cultivation has also been suggested despite minimal evidence.

There are various theories on the center of origin for the different millets (see Weber 1998 for a review). The millet plants relevant to this project are *Setaria italica*, *Panicum miliare*, and *Eleusine coracana*. An East Asian center of origin is suggested for *Setaria italica* by Werth (1937) and Rao et al. (1987). A South Asian and an East Asian center have both been suggested for *Panicum miliare* (Chalam and Venkateswarlu 1965; Ho 1975; Werth 1937; Vavilov 1959). And lastly, an African and a South Asian center of origin have been suggested for *Eleusine coracana* (Dixit et al. 1987; Mehra 1963; Savitri and Vishnu-Mittre 1978; Werth 1937). It is probable that particular millets were introduced at different points throughout the Harappan Phase (Mature through Late) (see Weber 1998). For example, based on the study of archaeobotanical materials recovered from Rojdi, Weber (1989) argues that *Setaria italica* did not become a prominent cultigen until after 2000 B.C. Although his argument is based on a single site, such changing emphases on different crops over time are not inconceivable. This could be a result of changing demands by the community and its subsistence system. For example, certain crops might be preferred over others at one point for their higher yields of vegetative parts, which are vital for fodder, fuel or building materials.

In general, understanding the adoption of new domesticates and agricultural practices has been overshadowed by research into indigenous origins of agriculture. The reasons and rewards for adding a new crop plant into an existing subsistence system are less frequently discussed and the motivation for people to begin to grow domesticates introduced from elsewhere remains an intriguing research topic. The patterns and processes leading to the introduction and the adoption of these cultivars are of great significance, although addressed infrequently. It is conceivable that *Setaria italica* was brought into South Asia by the seasonally migrating pastoral groups traveling the area between east-central Asia and Gujarat. The pivotal question in such a scenario is why would it have been beneficial for the South Asian populations to adopt these new cultivars? And additionally, why was the cultivar not adopted in areas in between the two regions?

Deciphering the Problem

As indicated in the proceeding review, the precise nature of Harappan settlement and subsistence patterns in Gujarat, although subject to considerable research, is yet to be fully defined. Investigations have focused on much needed settlement pattern surveys and site excavations, but there has been a general neglect of subsistence studies. In this context of an interplay between subsistence systems and settlement patterns (Bhan 1992; Possehl 1992), the economic importance and role of millets (drought resistant summer crops) at pastoral and agricultural settlements during the Mature and Late Harappan is relatively enigmatic and needs to be elucidated (yet see Weber 1989; Reddy 1991a). If indeed there was a deliberate shift from cultivation to pastoralism during the Late Harappan as argued by Rissman (1985), it is imperative to address this issue at both types of settlements: pastoral and farming. It is also important to model for possible interrelationships in subsistence between the two different categories of settlements proposed by Bhan (1989) (if indeed his categorization of the Gujarat Harappan is accepted) and by Possehl (1992) for the Sorath Harappan (small sites of migratory herdsmen and large permanent villages).

This research project has been designed to examine the role of millet crops at two sites (a very late Mature Harappan site and a Late Harappan site), and explore whether millet cultivation in Gujarat supplemented existing non-farming subsistence practices such as pastoralism. Thus, the elucidation of the role of millets, if any, in the increased emphasis on animal husbandry (or integration of agriculture and herding) is a major objective of this research. In particular, the use of millet byproducts or millet grain as fodder in 'pastoral' settlements and sedentary farming settlements is a key focus. Due to the overwhelming lack of consensus in terminology and chronology, my research at Babar Kot and Oriyo Timbo has focused on elucidating plant usage and subsistence orientation within a broad context of deurbanization. It has, however, tried to remain relatively nonpartisan in the Sorath/Sindhi Harappan distinction and cultural association. Of importance to the issues at hand is the subsistence system at play including deciphering the range of agriculture and pastoral adaptations at particular sites. In addition, emphasis is placed on identifying the potential range of adaptations between agriculture and pastoralism during the late Mature and into the Late Harappan to address potentially changing subsistence regimes.

The objective of this study is to further refine the economic role and importance of summer crops, specifically drought resistant millets, during the Harappan Phase in Gujarat. It is hypothesized that millet cultivation may have supplemented existing non-farming subsistence practices such as pastoralism, and played a varied role in an environment of changing subsistence and socio-political systems. Conceivably, millets may have been cultivated on a small level by the Harappan pastoral groups as animal fodder (Reddy 1991a). Millets have historically been very attractive fodders. It is plausible that use as animal fodder was the incentive for the initial South Asian interest in a plant such as *Setaria italica*. By investigating whether millets were used as fodder in the late Mature and into Late Harappan, new models of explanation can be developed to address the process of millet adoption in northwest India (Reddy 1991a). The objective here is to elucidate the importance of millets (and their cultivation if any) at pastoral communities and farming communities. In this light the economic role and use of these summer crops are examined and defined at a very late Mature Harappan site (Babar Kot) and a Late Harappan site (Oriyo Timbo) - potentially two very distinct sites in terms of subsistence strategies.

Based on ethnographic and ethnobotanical observations of present day Gujarat, the relationship between pastoral and agricultural subsistence strategies could have been quite complex. The most detailed and rigorous examination of the range of adaptations and interactions between pastoralism and agriculture has been conducted by Khazanov (1984) in Central Asia and has applicability to this region of India. Khazanov's (1984) definitions of the various pastoral nomadic adaptations are based on the degree of dependence on agriculture by pastoralists, and include five broad categories: pastoral nomadism; semi-nomadic pastoralism; semi-sedentary pastoralism; herdsman husbandry; sedentary animal husbandry. Pastoral nomadic adaptation is distinguished by the complete absence of agriculture. Semi-nomadic pastoralism is characterized by pastoralism being the dominant economic activity, with agriculture often playing a secondary, supplementary role. The two economic activities could be practiced (often on a seasonal basis) by different segments of the same group or not. Semi-sedentary pastoralism is distinguished from these two pastoral adaptations, since agriculture is the dominant economic activity, but seasonal pastoral activities are also important economically. Again, certain segments of the group may participate in these pastoral activities on a seasonal basis. Herdsman husbandry, or distant-pastures husbandry, entails pastoral activities pursued by specialists with the majority of the group being sedentary agriculturalists. The final form of pastoralism, sedentary animal husbandry, only functions to supplement agriculture, which is the predominant economic practice. Khazanov (1984) emphasizes that there are many variants in each of these basic forms of pastoralism. The differences between each form, however, have profound implications that need to be elucidated to fully understand the Late Harappan cultural, socio-economic and political changes. If it can be demonstrated that millets were cultivated for animal fodder by pastoral groups then the resulting subsistence system is more akin to semi-sedentary pastoralism, semi-nomadic pastoralism or herdsman husbandry, and not pastoral nomadism as proposed by Rissman (1985) and Bhan (1989), since fodder cultivation is the hallmark of a highly integrated system of agriculture and herding (Khazanov 1984).

In order to rigorously examine the precise role of millets in complex subsistence systems, new methodologies are needed. A methodology needs to be developed geared to elucidate these varied roles of millet cultivation, and to identify crop cultivation in archaeological contexts. Despite the research done to date, cultivation of the millets as an economic activity has not been identified at any Harappan sites in Gujarat. The simple recovery/ retrieval of archaeobotanical millet seeds from the sites is not adequate to argue for cultivation, since the grains could have been obtained from outside the site by the occupants. Identification of the cultivation practice is important to minimize concern that millet grains were being obtained by the Gujarat Harappans of a particular site through trade from elsewhere (from settlements within and/ or outside the region). Such research can be accomplished through the application of ethnographic models as done by Hillman (1984) and Jones (1987) for wheat and barley in the Near East and Greece, and for millet cultivation in the Late Harappan in Gujarat (see Reddy 1991a and 1991b).

Thus, this study has been designed to test several hypotheses regarding the role of millets in the Harappan subsistence economies, and to develop a methodology to identify "crop cultivation" in the archaeobotanical record. The first step was to design and conduct an ethnoarchaeological project aimed at modeling crop processing for archaeological application (chapters 3 and 4). Ethnographic observations particularly designed for archaeobotanical interpretations, as done here, have never been done in a South Asian context, or for millet crops. The results will therefore contribute to the methodology of agricultural ethnoarchaeology, and research design in the fields of archaeobotany and ethnobotany. The study also aims to provide new approaches, and offer innovative investigatory methods to facilitate gaining a better understanding of the subsistence and agricultural systems of Harappan Phase in Gujarat.

Ethnographic studies of crop processing activities have been used to make inferences about the archaeological use of specific crops (Dennell 1972, 1974, 1976; Hillman 1984; Jones 1983, 1987), and to build 'cause and effect' models for archaeological interpretation. The principle behind these studies is that a crop is processed through a number of stages before it is consumed, and that each step of crop husbandry and grain processing has a measurable effect on the composition of crop products and by-products. Depending on the type of crop, there are a set number of stages that it passes through. For example, all crops have to be threshed and winnowed, and the number of threshings and winnowings varies, as do the types of threshing and winnowing. Hillman (1981) argues that there are relatively few efficient methods of processing a crop available to non-mechanized farmers; in other words, a crop can be processed in only a limited number of ways. Hillman (1984) and Jones (1983, 1984) have also argued that husbandry practices (such as the choice of soil, tilling methods, sowing time, and irrigation) have a distinct effect on what weeds grow in the cultivated fields, and these processes are potentially detectable through archaeobotanical analysis of the weed seeds. Hillman (1981) and Jones (1983, 1984) define specific stages and contexts in the processing sequence of wheat, barley, and legumes through flow charts generated from studies of processing, disposal, and accidental charring. They conclude that specific ethnographic sample compositions result from a particular crop processing stage. Thus, given an archaeological sample of plant remains similar to documented ethnographic samples, it is inferred that these archaeological remains were the product of essentially similar crop processing methods. Predicted stages then can be matched to similar contexts in the archaeological record on the basis of plant frequencies. So, with adequate sampling for macrobotanical remains, investigators can identify the past activity at particular locations (Hillman 1981).

There are limitations in the previous predictive modeling studies (Hillman 1984; Jones 1984), since they are applicable only to wheat and barley. No such models have been developed for millet cultivation. Application of existing models to other crops (such as millets or corn) is not possible since the basic premises of the models is specific to the morphology of wheat and barley crops and the necessary processing methods and sequences. Additionally, the ethnographic studies of Hillman (1984) and Jones (1984) were in Turkey and Greece, both very different environmentally and culturally from Gujarat, India, and the crop processing stages of India are known to be different from the Near East. Therefore, it was necessary to conduct a pioneering ethnographic study of millet crop processing, to enable modeling for archaeological contexts in India.

The main limitation of such ethnographic studies is one of inference. To make statements about past crop processing and crop husbandry based on the charred archaeobotanical remains, a link between activities and the material observable must be made. This can be done by observing activities in the present that result in specific archaeologically visible patterns. There are, undoubtedly, problems with analogies between the past and present, since similarity between the present and the past in one respect or context does not necessarily imply it in another (Binford 1981; Hodder 1982). Therefore, methods should be sought at the analytical level to identify patterns in the data and evaluate their significance. Then interpretive theory comes into play in assigning meaning to these patterns (using ethnographic modeling), and subsequently evaluating the relative merits of a range of possible interpretations. The direct analogy problem is evident in uniformitarian arguments about people, however, in ethnographic studies of crop processing, such assumptions are made about plants and their behavior under defined physical conditions and restraints. Such a relationship is causal and relevant (Binford 1981), and, using Hodder's (1982) terms, is 'relational' rather than 'formal'. In any case, it is crucial that analogies take account of the context in which they are applicable (Hillman 1981; Hodder 1982).

The ethnographic crop processing studies of this study were not conducted in isolation from the works of Hillman (1984) and Jones (1984); instead, they were developed and structured to complement previous studies. The main objective of my ethnographic study was to discern the millet crop processing sequences, isolate variables that distinguish different stages, and to model for archaeological interpretations. An additional objective was to identify whether millets are processed differently depending on whether they are grown only as human food, both as human food and animal fodder, or when grown only as fodder. The principle behind these studies is that any distinction in processing (for the three different purposes) would have a measurable effect on the composition of crop products and byproducts. It was therefore necessary to isolate the variables which distinguish crop processing stages when millets can be grown for the three different purposes: food, fodder, or both. My detailed ethnographic research was designed to quantitatively discriminate the variables that correlate the cultivation of millets with their ultimate use. The processing pathways of summer crops grown for different purposes were examined and samples collected from each processing stage for analysis. The models generated were then testable archaeologically, if a sufficient sample of analogous plant materials were recovered through flotation.

Two main ethnographic studies of millet crop processing in India were conducted: summer cultivation and 'Opportunistic' cultivation. Summer cultivation of millets (*Pennisetum typhoides* and *Sorghum bicolor*) was studied in Gujarat, India, which is also the locale of the two Harappan sites, Babar Kot and Oriyo Timbo to which this study was applied. Traditional cultivation and processing of millet crops is still practiced in this region, and Gujarat offered the ideal venue to conduct such a study of crop husbandry and processing. The other ethnographic study involved the exploitation of a flood plain in South India through a specific form of agricultural practice termed here as 'Opportunistic' cultivation- an extensive form of cultivation exploiting the annual flooding of the river. Extensive 'Opportunistic' cultivation also could be an important economic activity elsewhere (such as prehistoric Egypt and Europe), where ancient civilizations focused along the river courses. The study of this form of cultivation for archaeological interpretations is scarce but very important. From time immemorial river flood plains have been an attractive environment for human exploitation, particularly for agricultural and pastoral pursuits. Well-known examples of such occupancy, prehistoric and historic, range from settlement occupation along the Mississippi and the Amazon in the New World, to the Danube, Nile, Tigris, Euphrates, Indus and Huang Ho in the Old World. River flood plains are dynamic and each is unique in its nature. These riverine systems have been manipulated for agricultural activities and quite often human settlement patterns have been strongly influenced by these river systems and their flood plains. The Indus civilization was set around two major river systems and their flood plains provided vast areas for grazing and agriculture. The diversity of the subsistence base and resource variability with the potential for surplus production was one of the preconditions needed for the successful development of integration and urbanization of the area (Kenoyer 1991). The fertile flood plains of the region were highly productive without massive irrigation systems. In the light of this, the study of the 'Opportunistic' cultivation and the understanding of the dynamics of its economy are valuable for furthering our understanding of the growth of the Indus Valley Tradition.

An Integrated Approach

To recapitulate the goals and hypotheses of this endeavor, the primary objective was to define the economic role and importance of summer crops, specifically drought resistant millets, during the late Mature and into Late Harappan in Gujarat. I have hypothesized that millet cultivation may have supplemented existing non-farming subsistence practices such as pastoralism, and had a varied role in an environment of changing subsistence and socio-political systems. Of particular importance is the elucidation of two issues: (1) whether millet crops were cultivated by the occupants of Oriyo Timbo and Babar Kot, or were they traded and brought into the settlement; (2) the ultimate use of the millets (regardless of whether they are imported or cultivated) either as human food, animal fodder, or a combination of both. To successfully address these research questions, it is essential to integrate several avenues of investigation, including developing new methodological framework based on ethnoarchaeological models of crop processing.

Since the current understanding of Harappan subsistence regime in Gujarat is primarily conjectural, the use of multiple lines of evidence has considerable potential to strongly contribute to understanding of the processes and character of the Harappan subsistence systems. Paleoethnobotanical investigations are certainly necessary to address these research questions. However, given paleoethnobotanical limitations and the need to address predepositional and depositional aspects of middle range research, ethnographic crop processing studies were also needed. Thus, integrating these two complementing approaches to subsistence studies provides for robust and more detailed interpretations of the archaeobotanical data.

Crop processing studies and paleoethnobotany, however, cannot with absolute certainty resolve the question of whether millets were used as fodder. Carbon isotope analysis of animal bones was chosen in this project to complement these two approaches. The basis of carbon isotope analysis rests on the scientific determinate that there are three types of photosynthetic pathways in plants (C3, C4, and CAM) and there are two stable carbon isotopes in the biosphere that are taken up differentially by all living creatures (Carbon 12 and Carbon 13). The ratios of these two stable isotopes of carbon in animal bone collagen have been used to determine the relative contribution of C4 plants (such as maize and millets) versus C3 plants (such as wheat and barley) to the diet of the consumer (Deniro and Epstein 1978; van der Merwe 1982). If millets were used as a major source of animal fodder, then the higher proportion of C4 plants in the animal diet would be reflected in the carbon isotope signature of the animal bone collagen. Thus, carbon isotope analysis can be used to determine the use of millets/C4 plants as animal fodder. This analysis cannot distinguish, however, between the use of millet grains versus millet crop residues as animal fodder. However, this distinction is important because the use of millet grain for animal fodder would indicate that the crops were primarily grown for animals, whereas using only the millet residues for fodder implies that the crops were grown for both human and animal consumption. These differences significantly affect interpretation of past subsistence economies. Crop processing studies provide a viable complementary line of evidence to identify this distinction.

For a thorough and comprehensive investigation of the problems related to food and fodder, the quantitative results from the three analytical methods need to be evaluated and integrated. For example, since variables related to seasonality and differential feeding of grains versus crop by-products cannot be identified and defined by carbon isotope analysis, and their elucidation is essential for the interpretation of past economies, it is necessary to complement carbon isotope analysis with other lines of evidence. Thus, this study intertwines the three lines of evidence (ethnographic studies of crop processing, paleoethnobotany, and carbon isotope analysis) to offer robust and new interpretations of the Harappan subsistence practices.

3. The Living Past: Ethnographic Crop Processing Studies

A sound methodology linking archaeologically recovered plant remains to past human agricultural activities is essential for the study of agricultural practices and subsistence systems. This chapter presents an account of the ethnoarchaeological fieldwork and observations related to crop processing activities. It is structured to address summer cultivation and winter 'Opportunistic' cultivation of millets, as case studies for archaeological interpretation. Summer cultivation is presented first, with a detailed discussion of the crop processing activities relevant to this study. The second half of the chapter examines winter 'Opportunistic' cultivation and its associated crop processing activities. The emphasis of the discussion is on archaeologically relevant and important contexts, activities, and processes.

The two geographically distinct study areas in India were selected to facilitate examination of seasonal and geographical variations in millet crop processing (Figure 3-1). In order to successfully model for a range of archaeological situations and to build predictive models that can be adjusted and applied globally, ethnographic data should not be too specific. Therefore, it was advantageous to collect data from two geographically and seasonally different loci: winter flood plain cultivation in the modern state of Andhra Pradesh (South India) and summer monsoon cultivation in the modern state of Gujarat (Western India) (Figure 3-1). Three factors determined the selection of these particular study areas. The two specific agricultural systems studied were selected because they presented ideal logistics for research and study. In addition, the

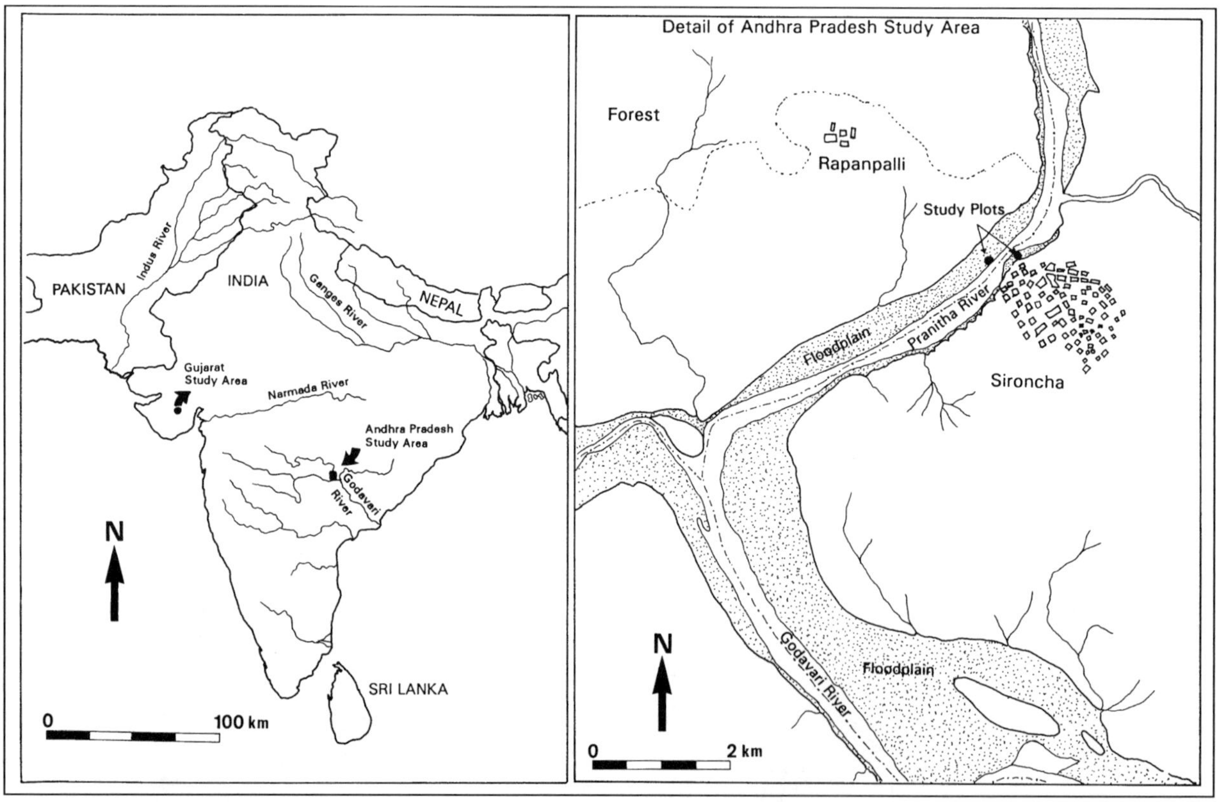

Fig. 3-1. Ethnographic crop processing study areas.

two farming systems are significantly different in terms of economic strategy, yet they both have relevance to Harappan subsistence economy. Winter 'Opportunistic' cultivation is an extensive form of economy, while the summer monsoonal cultivation is an intensive economic activity. Potentially both systems of subsistence procurement could have had related counterparts in the Harappan times; for example 'Opportunistic' cultivation along all the major river systems and summer cultivation in Gujarat and Punjab.

The third and most important factor that determined the selection of the two study areas entails the types of crops involved. Winter 'Opportunistic' cultivation involved the cultivation of the millet *Panicum miliare*, while summer cultivation primarily involved the millets *Sorghum bicolor* and *Pennisetum typhoides*. These crops are distinct in their morphological characteristics relevant to crop processing, specifically in terms of harvesting methods. It is proposed here that the morphology of the crop plant determines the type of harvest technique that can be used. In plant crops such as *Sorghum bicolor, Pennisetum typhoides* and *Eleucine coracana*, the stalks/stems are thick and only one to three plants can be cut at a time. This automatically provides for selection against most weeds being included in the harvested materials, because when only one to three stalks are being cut, the harvester can select the crop plant stalks for harvest. In addition, the seed bearing 'heads' (panicles) of these plants are compact and larger in size. Therefore, often after harvest they are separated from the stalks and only these seed bearing heads are processed subsequently. Furthermore, harvesting can also be done by cutting off only the panicles (the seed bearing earheads). This harvesting procedure leaves the rest of the plant standing, and also selects out the weeds. These crops have been grouped as Type A crops (Reddy 1991a, 1991c), and the harvest method termed Type I harvest (Figure 3-2).

In contrast, crops such as *Panicum miliare* and *Setaria* spp. do not have compact and distinct seed bearing panicles, and their stalks are relatively thin. As a result of the slender stalks/stem, these plants are harvested by gathering a large number of plants together while cutting (Figure 3-3). This makes selection against weeds, growing in the field among the crops, difficult and often too time consuming. Therefore, weeds get incorporated into the harvested materials and then processed out in the subsequent stages of processing. The separation of seeds from the plant therefore requires the harvest of the entire plant and the subsequent separation of crop grain through various processes. These crops are grouped as Type B crops and their harvest technique is termed Type II harvest (Figure 3-3).

Therefore, the critical difference between the two crops and their harvest methods is in the inclusion of weeds into the harvested material. This is a significant variable because weeds play an important role as indicators of various processing stages (Hillman 1984; Jones 1987). In Type A crops harvested in method I, there are very few to almost no weeds in the subsequent stages, since they were not initially included in the harvested materials. In the second harvest method (Type II), the weeds are included at harvest because they cannot be selected out without considerable effort on the part of the harvester. Figure 3-4 shows this process and the ramifications of the harvesting techniques. These two distinct situations have great implications for archaeological contexts and interpretations related to crop processing. To model the processing pathways of both the crop types (Type A and Type B), the harvest methods of each were studied in Gujarat and Andhra Pradesh.

Summer Cultivation of Type A Crops (*Sorghum bicolor* and *Pennisetum typhoides*)

Geographical Setting and Economy of Study Area

A season of the ethnographic crop processing studies was conducted in northwest India, in the modern state of Gujarat, which is approximately coterminous with the southern region of Harappan culture (Possehl 1980:18). The study area is located in a peninsular region called Saurashtra, an area rich in Harappan sites of varying sizes and types. The ethnographic study was done in the vicinity of the Harappan site of Babar Kot. The soils in the area are black cotton soils, which have high water retaining capacity. The rainfall is low in the area, falling almost exclusively in the monsoon seasons. An average of 60 cm of precipitation per year is common in Gujarat; the monsoons of June - September are often poor, thereby bringing very little moisture. The ethnographic study was done in the months of September through December, since the harvest season spans from late October through November.

Land use in the study area is similar to the rest of the modern state of Gujarat, and typical of semi-arid India. Dry farming is the most common subsistence activity in this region. The major agricultural season is the 'kharif' monsoon season, which involves sowing with the first monsoonal showers in the months of June or July, and harvesting in October-November. Winter cultivation occurs only with well irrigation, and on selected farmsteads. The crops cultivated during the two agricultural seasons are distinct. 'Millets' which are drought resistant crops are cultivated in the monsoonal season, while wheat is cultivated in the winter with the aid of irrigation. Occasionally *Sorghum bicolor* and *Pennisetum typhoides* are cultivated in the winter months with the use of irrigation. However, it is most often the case that these millets are hybrids or/and cultivated for green fodder. In recent times there has been a persistent warm drought condition in Saurashtra, and even the winter months have been warm enough to successfully cultivate millets. The monsoons have failed repeatedly in very localized areas, and often there is minimal success of cultivation in these areas. Usually when the crops fail to seed well, they are ultimately used as fodder. Interviews with farmers were conducted to gather ethnohistorical information about whether this present pattern of rainfall

Fig. 3-2. Type I harvest method of Type A crops.

possibly occurred earlier. These farmers strongly assert that the failure of monsoons is only an event of the late 20th century.

Problem Orientation and Field Methodology

In this study, summer cultivation of millets in Gujarat was addressed to: provide the regional perspective on agricultural systems and crop processing studies for the Harappan period; study the cultivation in a river plain where there are weeds comparable to archaeological situations; study the crop processing of Type A crops such as *Sorghum bicolor* and *Pennisetum typhoides*, and lastly to have geographical and seasonal variation. The specific location of the study area was selected because it is in close proximity to the Harappan site of Babar Kot, studied for this project, which facilitates archaeological interpretations since the soil, geology, and water variables are comparable. On the basis of extant, admittedly scant evidence, scholars have reasonably argued that the development in South Asia of summer, rain fed, 'millet'-based agriculture took place largely on the eastern and southern margins of the Indus Valley. This occurred during the Harappan period in areas where monsoon rainfall was sufficient to support the dry farming of these hardy, warmth-loving cereals. Therefore, Gujarat provided the ideal area to conduct ethnographic studies due to the Harappan tradition present in the region and the specific geographical location, thereby ensuring that the ethnographic study area and relevant archeological sites are in a similar environmental zone.

Fig. 3-3. Type II harvest method of Type B crops.

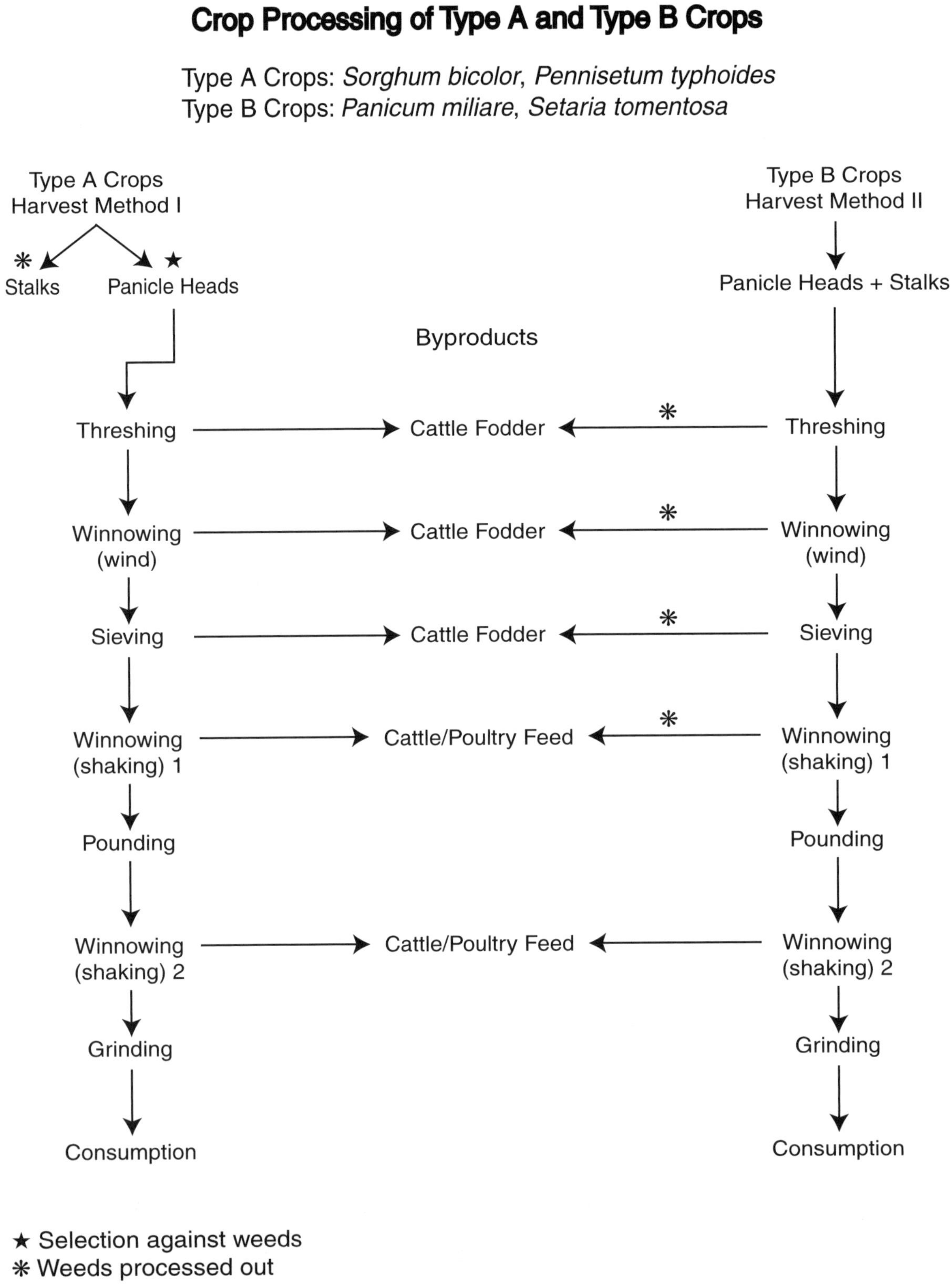

Fig. 3-4. Crop processing pathways: Type A and Type B Crops.

The modern agriculture of Gujarat is characterized by a reliance upon dry farming with eighty five percent of the agriculture being rain fed (Patel 1977). The great majority of crop production occurs in the kharif (summer) monsoon season, in contrast to rabi (winter) production. The main crops grown in the area are *Pennisetum typhoides* and *Sorghum bicolor*. *Pennisetum typhoides* is cultivated under more acreage than any other crop, mainly because of its short maturation period. *Sorghum bicolor* however, replaces *Pennisetum typhoides* as the most popular food grain in South Gujarat. It is also an excellent fodder crop for cattle (Patel 1977). Aspects of cultivation such as land preparation, sowing pattern, plant spacing, harvest timing, harvesting methods, crop processing at the field and in houses, and storage areas were studied to identify distinctive variables at each step.

The ethnographic crop processing fieldwork of summer cultivation took place in the settled village of Babar Kot in the Bhavnagar District of the modern state of Gujarat (see Figure 3-1). The farmer, whose farm and crops were studied, practiced traditional agriculture and mechanized equipment was not used for processing the crops grown on his land. Two millet crops were chosen for study, those of *Sorghum bicolor* (Jowar) and *Pennisetum typhoides* (Bajri). These crops were chosen to complement the study of *Panicum miliare* in Andhra Pradesh since the former crops are classified as Type A crops while the latter is a Type B crop (Reddy 1991c). The selection of the crops was also determined by the availability of such fields for study. Based on ethnographic studies in Greece, Jones (1987:44) suggested that there are 14 processing stages for free-threshing cereals and pulses. This project was designed to examine whether these 14 stages or other stages are applicable to millet processing. During the fieldwork, emphasis was particularly placed on documenting the cultivation of millets and their ultimate use, and the research was designed to quantitatively discriminate the variables that correlate the cultivation and processing of millets with their ultimate use.

Before the harvest, interviews were conducted over a series of days. These interviews with the farmers included gathering information regarding the crop cultivation, selection of crops for cultivation, preparation of fields, details of the actual process of sowing for different crops during different seasons, and so on. Upon completion of the interviewing, different weeds associated with the millet fields were collected. The collection was facilitated by the farmer, whose knowledge of such associations was extensive. It was important to collect these weeds, in order to have a comparative collection during the analysis of the harvested samples of products and byproducts. It is assumed that the weeds associated with the millet fields will be the ones that will be indicative of different products and byproducts of different stages of crop processing. Such weeds were also collected the previous year and processed for their carbon isotopic signatures to provide insight on the nature of the environment. The majority of the weeds were C3 plants, which will not confuse the C4 signature of the millets.

After the weeds in the millet fields were compiled into a comparative collection in September before the crop harvest, study plots were selected in the fields of *Pennisetum typhoides* and *Sorghum bicolor*. Three plots, measuring 5 by 5 m, were selected in each of these two crop fields. The selection of the plots was based on the following criteria:

1) Distance to water source: closer and further
2) Distance to cattle shed: closer and further
3) Distance from field of another crop: closer and further
4) Degree of isolation within the field: high and low
5) Proximity to a pathway: closer and further

These above criteria may significantly affect the ratio of crop plants versus weed plants, and controlling for these variables will be critical during the analysis and interpretation. Between the three study plots, all the variables of the five criteria were fulfilled. The 5 by 5 meter squares were laid out in the field using triangulation (Figure 3-5). Long stakes were used to demarcate the study plots, because of the plants' height. A pace map was made showing all the activity areas related to cattle keeping, goat keeping, domestic area, field areas of various crops, and weed patches. Irrigation channels, when present, were carefully mapped.

Before and after each of these processing stages, weights of the products and byproducts were recorded, and then samples were taken from each. An attempt was made to take a minimum of 15% of the weight for a sample. While taking the sample, careful attention was paid to take a homogeneous sample representative of the whole product or byproduct. The uses of each byproduct were noted as the crop processing was being done.

Of the three plots laid in the *Sorghum bicolor* field, one plot was very weedy, the predominant weed being *Setaria tomentosa*. This plot was stud-

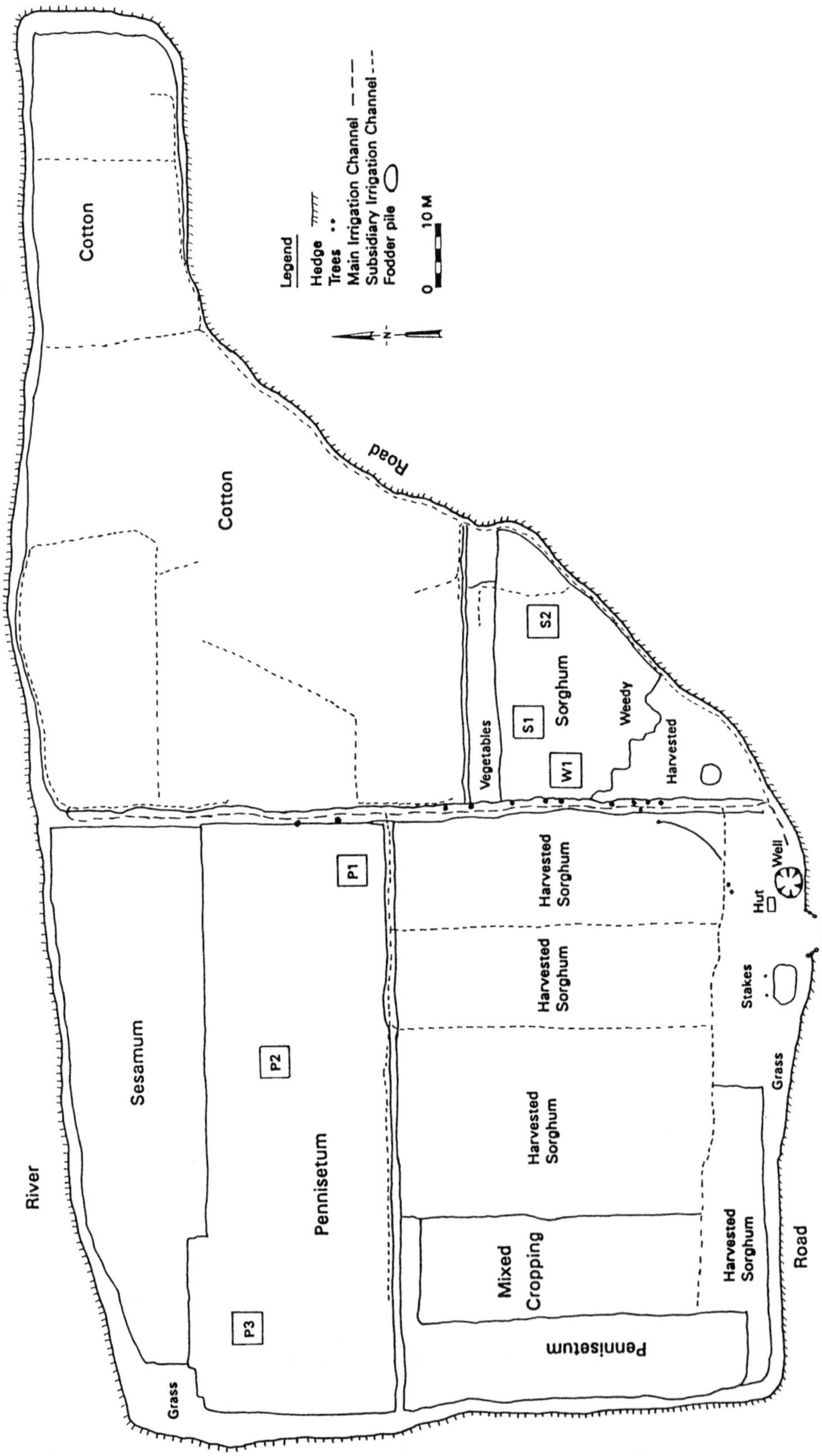

Fig. 3-5. Summer cultivation study plots and the Babar Kot farm.

ied as a distinct plot, since it can be used as an example of wild plant processing for food and/or fodder. The processing done for this plot was similar to that done to *Sorghum bicolor* and *Pennisetum typhoides*, with *Setaria tomentosa* being selected as the final end product. It is a small seeded annual grass that grows wild in Gujarat and is an attractive animal fodder. The study plot of this plant was at the edge of the *Sorghum bicolor* field where the weed grew wild and took over the area since there were very few *Sorghum bicolor* plants.

Botany of the Crops

Both *Pennisetum typhoides* and *Sorghum bicolor* are tropical grasses and their seeds are similar in that they can be viewed as 'naked' seeds. They are not enclosed in a hard binding seed husk and this has a significant quantifiable difference with respect to processing of the grain for consumption. *Pennisetum typhoides* a robust annual grass about 1.5 to 3 m tall with solid stems about 1 to 2 cm in diameter (Rachie and Majmudar 1980) (Figure 3-6). The leaves are long, slender, and hairy. The plant has a deep root system, which helps it survive in drought situations. The seeds are borne on inflorescence heads that are compact, unbranched, and cylindrical. The rachis is straight and about 8-9 mm thick, and bears the spikelets on an involucre and rachillas. Usually there are an average of 1600 spikelets per panicle head. Each spikelet consists of a short lower glume and an inner longer one. It also has two flowers, one of which is fertile, and occasionally both are fertile and seed bearing. The lemma and palea of the flower do not clasp the caryopsis, which threshes free (Rachie and Majmudar 1980). The caryopsis or seeds are large in size. The elements of the plant that occur in the different crop processing stages are the inflorescence heads with and without grain, rachis fragments, spikelets with and without seeds, grains with glumes, and grains without glumes. *Pennisetum typhoides* is mostly grown as a rain fed monsoon weather crop, however to a limited extent it is also grown as an irrigated hot weather crop in central and south India.

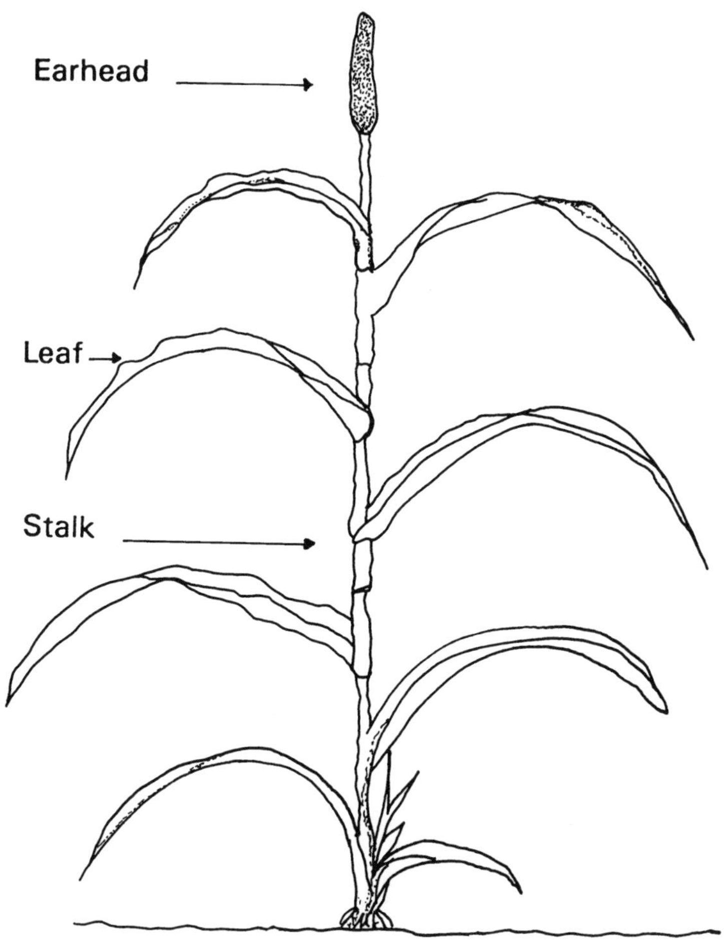

Fig. 3-6. *Pennisetum typhoides* plant.

The *Sorghum bicolor* crop plant is similar in that the plant is an annual, and has an erect solid stem supported by a profusely branched adventitious root system (Figure 3-7). The stem bears the compact inflorescence head at the tip. It is a more or less compact panicle with an axis or rachis of variable length (Cobley 1956). The spikelets of *Sorghum bicolor* occur in pairs on the rachis on the tips of branched stiff branchlets. Of the two spikelets the sessile one produces the grain. It has two glumes and contains two flowers of which only one bears the seed. The maturity of the head starts at the apex of the cylindrical cone and continues down. The caryopsis is rounded and varies in shape and size, and it comes lose of the palea and lemma without much threshing. The plant parts and seed parts that are recovered in the processing stages are similar to those recovered for *Pennisetum typhoides*.

Sowing

Sowing of both *Pennisetum typhoides* and *Sorghum bicolor* is done in the monsoon month of June during the rains. For a week before sowing, the field is ploughed with a wooden plough using two bullocks. Dung is often mixed into the field at this time. There is no irrigation during sowing, since it is done during the rainy season. Using two bullocks with a three-pipe structure attached to them

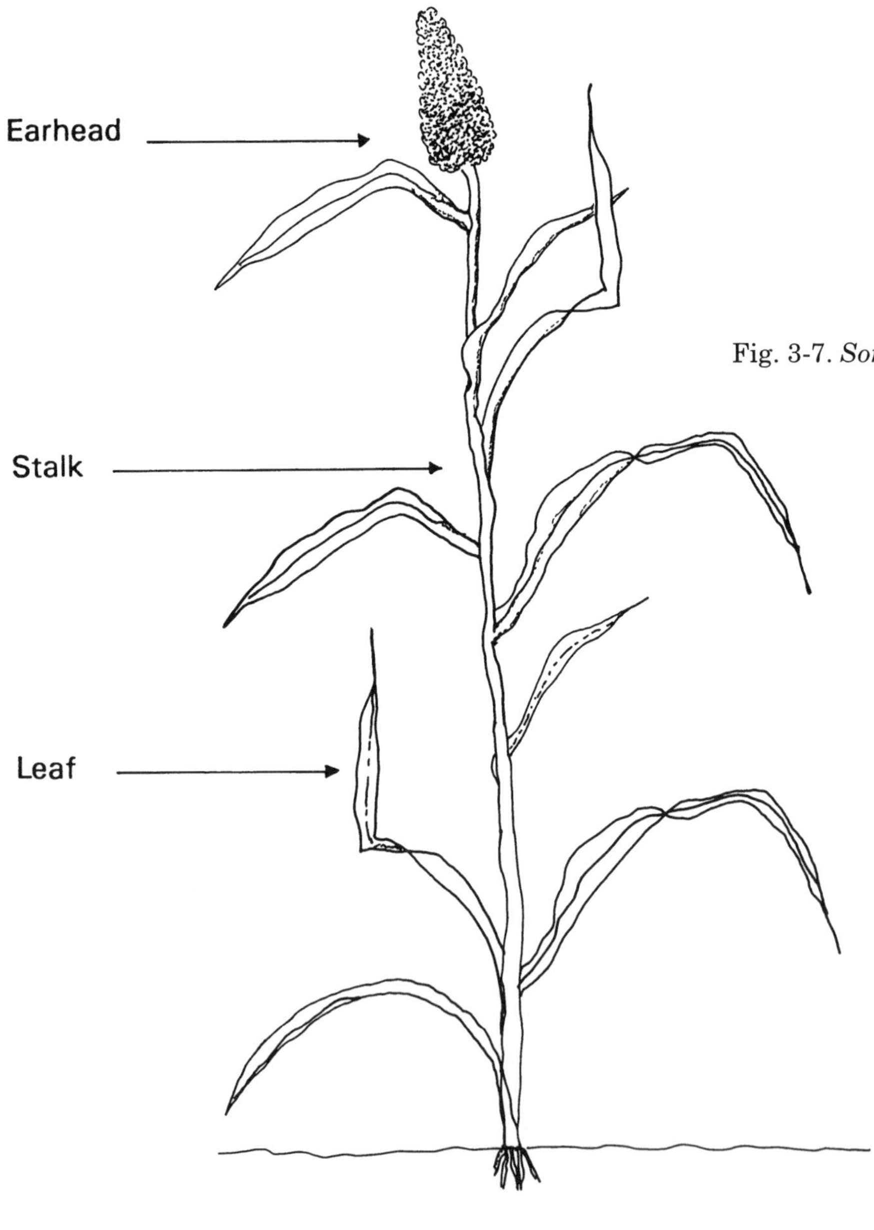

Fig. 3-7. *Sorghum bicolor* plant.

like a plough, sowing is done as a one-day event for each field. The seeds drop out of the three pipes, and the spacing of the plants is about 6-8 inches apart. If the spacing is any closer, the inflorescence head is quite small. Ordinarily, irrigation is not needed if the monsoons are good, but in recent decades there has been a decrease in the quantity of rainfall and success of the monsoons. Therefore, farmers are often forced to use well irrigation about once every five days. When there is no well available then obviously no irrigation is done. The maturation of these crops is simultaneous, because the sowing is a one or two day event. When the land holding is large, then the sowing is done in lots so that the harvest times do not coincide too much. This, however, was not an issue at the farm studied, which was approximately 5.5 acres in size, and considered a small farm.

Harvesting

Harvesting is the reaping and gathering of grain or other products from the plant. This can occur in two ways: uprooting of the plant (where actually there is no separation of the component such as a panicle from the rest of the plant), and the cutting of the grain bearing part from the vegetative parts. Uprooting is less common in India as it is often more labor intensive in the subsequent stages of crop processing. It is more common when the crop concerned is being grown for fodder, and the subsequent stages are irrelevant. The second method of harvesting, cutting, is the most common method.

The harvesting and crop processing of the millet plants from the study plots in Gujarat was conducted by the farmers. *Sorghum bicolor* and *Pennisetum typhoides* are Type A crops (Figures 3-2; 3-4), therefore harvesting is most commonly done by selecting out the weeds while cutting, since only a few individual plants are cut at one time. When labor is available in cases of large family holdings or richer landowners, often enough labor is employed to first separate the heads of the crops from the stalk in the field itself. The inflorescence heads are processed for grain, and then later the stalks are cut down for fodder (Figure 3-8). In this case there is very little weed contamination in the crop processing, since they are selected out in the first step, and the crop processing involves only the separation of the seed from the other crop parts such as chaff, straw, and glumes. This process does involve considerable labor and the number of hours invested in the harvesting process is more since it is a two step process, and often only worthwhile in cases of large land holdings with surplus labor. It is, however, noteworthy for archaeological interpretations.

In most cases, the crop plants are cut by selecting one to three plants at a time and then carried to the threshing floor where the heads are separated from the stalks before threshing (Figure 3-9). The harvested plants are stacked in a row in the field, and subsequently when a sizable bundle is formed they are taken to the threshing floor. In such a harvest, the weeds included in the threshing material are limited and comprise of only those that were harvested at a time with the selected one to three crop plants. This is significantly different from Type B crop harvesting.

Since it was not possible to observe a Type B crop that had weeds included in the harvested material (the 'Opportunistic' cultivation crop was sterile for any weeds), it was important to accommodate this need. *Pennisetum typhoides* and *Sorghum bicolor* crops were harvested to simulate Type B crops while taking into account the processes that a Type A crop would be undergoing. So, *Pennisetum typhoides* and *Sorghum bicolor* were harvested by cutting at the base of the stalk, including all the weeds in the plot, as it would be if the crop were a Type B crop such as *Panicum miliare*. While this was being done, the way *Pennisetum typhoides* and *Sorghum bicolor* would normally be harvested as a Type A crop was also noted. This would have been by selecting only one to three plants at a time and thereby decreasing the number and types of weeds included in the harvested materials. These observations were made by noting which weeds were tall enough to be included in a *Pennisetum typhoides* and *Sorghum bicolor* harvest, since only the taller ones growing close to the bases of the stalks would get included in this harvest.

A distinctive characteristic of *Pennisetum typhoides* is that it seldom matures uniformly which in turn implies harvesting in stages, or by 'picking'. Rachie and Majmudar (1980) report two methods of harvest in unmechanized agriculture. First involves the cutting of heads (as they mature) and placing them in the sun to dry, while the second involves the harvesting of the entire plants, removing the heads with a sickle or knife after a few days of drying. They report that this latter method is less frequent and is used in two situations; when birds are not a problem, and when the soil is porous and dries up quickly during maturation (Rachie & Majmudar 1980). This could also be applicable to *Sorghum bicolor*.

Thus, it is important to note that Type A crops such as *Pennisetum typhoides*, *Sorghum bicolor*, and *Eleucine coracana* can be harvested in three different ways. The first situation is the harvesting of only the heads, as discussed earlier, where the inflorescence heads bearing the grain are first harvested separately by cutting them from the stalk (as shown in Figure 3-8). This results in clean harvest material where there are no weeds or stalks, only the inflorescence heads bearing grain. The resulting threshing and consequent processing stages have only crop grain, rachii, florets, glumes, and crop inflorescence heads without seeds as composition components. No weeds are present as byproduct components since they were not harvested initially with the harvest product. This type of harvest, though efficient in terms of cleaning involvement, is more labor intensive in terms of harvest time since the stalks are left in the field to be harvested again for fodder. This procedure is less common, and occurs only in select few cases where labor is not an issue. However, it cannot be ruled out for archaeological relevance. This is why it is important to isolate which inflorescence parts are distinguishing variables for different crop processing stages and pathways.

The second variation is when the inflorescence heads are harvested along with about 3/4 of the

Fig. 3-8. Labor intensive harvest method of Type A crop.

stalk. In this case some tall weed species are cut along with the stalk even when there is selection against the weeds by cutting of one to three crop plants at a time. This is the more common method, since it is only a one step process. Cattle are left in the fields to feed on the stubble left behind. The inflorescence heads are often separated from the stalks on the threshing floor, and at this time some of the harvested weeds get included in the threshing material. Therefore, the subsequent processing stages have weeds to be processed out in addition to the other inflorescence parts. This is similar to the inflorescence parts processing if the harvesting was done the first way. The stalks are fed to cattle as fresh fodder.

The third harvesting variation for *Pennisetum typhoides* and *Sorghum bicolor* is when heads, stalks, and all weeds are cut together. In this instance, weeds of all heights, except the prostrate types, would be cut and included with the harvested material, as would be the case in the harvest of Type B crops. The cutting would be at the base of the stalk, therefore, incorporating all the weeds that would not get included in the second type of harvest. This scenario is very rare ethnographically, but it is important not to dismiss it for archaeological applications. This is because it is possible that this type of harvesting could have occurred prehistorically when thicker stalk varieties of *Pennisetum typhoides* and *Sorghum bicolor* were not yet developed and it could be possible that with a thinner stalk this third type of harvest would have been the most feasible. Modern genetic manipulations have resulted in the selection and breeding for thicker stalks to provide more volume for fodder. It is therefore important to take into account all possible variations and study the resulting patterns for archaeological interpretation. It would then be possible to identify all these abovementioned variations of harvesting techniques archaeologically even though the techniques are quite homogeneous ethnographically. Ethnographically, the harvesting variation one and three are not as popular as variation two, but this might not be the case prehistorically, and the research design has taken into account this possibility.

During the study, it was imperative to note which one of these procedures will work for all three

Fig. 3-9. Post-harvest separation of Type A crop heads.

cases. It was decided that the third type of harvest method for *Pennisetum typhoides* and *Sorghum bicolor* would be most representative of all three types, since the patterns of processing of the other two could automatically be deduced from studying the processing of the third type of harvest. Therefore, it is possible to understand all three harvesting types by studying the third type. In addition, there will also be insights into Type B harvesting patterns, since the third type of harvesting of *Pennisetum typhoides* and *Sorghum bicolor* is similar to the harvesting of Type B crops.

It is important to note that this is not just ethnographic observation and study, but it is experimental ethnoarchaeology. This is because to learn certain types of information that are relevant prehistorically but not present ethnographically, a procedure was developed to recreate situations. This provides the means to understand the situation ethnographically and then model for prehistorical interpretations. We know the present is different from the past, and direct analogy is rarely applicable, and these kinds of ethno-experimental studies are necessary to avoid such errors common in direct analogy. This type of study is similar to growing plants and experimenting with them for domestication rates (Hillman and Davies 1990).

With the above in mind, the *Pennisetum typhoides* and *Sorghum bicolor* plots were harvested by cutting the crop at the base of the stalks and incorporating any weeds present. Traditional metal harvesting sickles were used for the cutting. The *Pennisetum typhoides* crop was harvested first and all the processing completed before the *Sorghum bicolor* crop and the *Setaria tomentosa* plots were harvested.

Crop Processing

In general (regardless of what crop is being processed) each crop processing stage produces two groups of assemblages: crop product and crop byproduct (or residue). The crop product (at each stage) is the selected assemblage for subsequent processes. The crop byproduct is the assemblage that has been processed out, and is that part of the harvest that is not included in the rest of the processing pathway. The compositions of these two assemblages are distinct from each other at each stage and at different stages.

The processing stages for the *Sorghum bicolor* and *Pennisetum typhoides* as Type A crops are different from the processing stages of Type B crops. The difference might not seem striking, but significant enough to make a statistically relevant variation. The *Sorghum bicolor* and *Pennisetum typhoides* crops are processed through roughly 11 stages after harvest and before they are ready for consumption. The following processing stages apply to *Pennisetum typhoides* and *Sorghum bicolor* (Type A crop processing).

Drying of the Crop Plants after Harvesting

The harvested crop plants are well dried in the sun before the next processing stage. This is done so the seeds will be dislodged with ease from the heads. If the heads are wet the dislodging is ineffective, and in addition, if the heads are wet, there is a possibility of later fungal infection, which destroys the crop produce.

Preparation of the Threshing Floor

A hardened floor was prepared at the edge of the fields for threshing. Its function is to facilitate efficient threshing of the seed bearing plant parts, to reduce the amount of soil that would enter the threshing product, and to reduce loss of seeds. The threshing floor preparation involved several steps over a four-day period. The selected area, approximately 5 m in diameter, was first cleaned of all ground cover. Then, it was watered down well to settle the dust and left to dry for a day while periodically watering it (Figure 3-10). On the second day, two basket loads of dung were spread out and watered down. Two bullocks were then walked over this slushy surface for 4 - 4.5 hours to harden it. After this the threshing floor was evened out by plastering by hand with dung and water. It was then left to dry overnight, and on the third day it was plastered again, and ready for use after it was well dried, which usually takes a day. It should be noted that these field threshing floors are subsequently ploughed and replanted, thereby obliterating them from the archaeological record.

Examination of a threshing floor sample after all the crop processing was completed revealed a compact surface encrusted with crop and weed seeds at varying stages of desiccation and integrity. The original source of these seeds is difficult to elucidate since they could be sourced to the dung used to plaster the threshing floor, and also to the crop processing that occurred on the floor. This context poses an interesting and also a potentially confusing issue for the archaeological interpretation of plant remains that are found embedded in threshing floors where dung plaster is used.

Threshing with Cattle

Threshing is the process by which the seed bearing parts of the plant are separated from the non-seed bearing parts. The material to be threshed is laid out on the threshing floor, and then the bullocks held together by the farmer are led to tread on these materials in a systematic manner (Figure 3-11). This activity is predominantly done by men, while related activities such as sweeping up and other assistance are done by women (Figure 3-12). The dispersal of material from the heap is periodically swept into a pile to facilitate efficient threshing. This threshing process for each 25-meter square, harvested plot took approximately 1.5 hours. The process results in a byproduct that is essentially made of inflorescence heads without seeds or with a minimal number of seeds, and small immature seeds. There are also weed seeds, and weed plants with and without seeds. This byproduct is fed to the cattle in the field. Occasionally it is taken home to be stored for later use. This later use is usually in about a month since *Pennisetum typhoides* and *Sorghum bicolor* stalks and chaff do not store well for long periods of time unless they are treated chemically.

Threshing with stick

After the threshing with the cattle is completed, there often remain some inflorescence heads with seeds that are left unthreshed. Such heads are selected out from the pile and threshed on by beating with a stick (Figure 3-13). This selective threshing is more time consuming, and is an indicator of extensive farming. This activity is done by both men and women depending on availability.

Sorting

This activity is done after the threshing stage. It is done by hand, and involves the physical separation of desired versus undesirable plant material from the product. Sorting is done by men and women in preparation for the next stage. It is difficult to quantify this process, since most of it is subjective and dependent on the sorter, the crop, and the time available for such an activity. Sorting

Fig. 3-10. Threshing floor preparation in Gujarat.

Fig. 3-11. Threshing with cattle in Gujarat.

Fig. 3-12. Sweeping up after threshing with cattle.

after threshing can also be done with the use of a rake, wherein the straw and other vegetative parts are raked out leaving the smaller material for the next stage. The rakes were introduced into the area by the British, and the present-day rakes look very much like the British Isles rakes. There is no distinct product or byproduct from the sorting process, since they are mixed into pre-existing groups of products and byproducts.

Winnowing (1)

This process involves the separation of grain from chaff, straw and other small light plant parts, and could occur on the threshing floor in the field. Winnowing can be done through two methods: winnowing by wind and winnowing by shaking. For Type A crops such as *Sorghum bicolor* and *Pennisetum typhoides* winnowing by wind is done first in the field followed by winnowing by shaking later in domestic contexts.

By wind

The number of times one winnowing method is performed on a given crop product varies depending on the product quality and quantity. Winnowing by wind is a common method used in most parts of the world where traditional agriculture is practiced (Figure 3-14). It essentially uses gentle wind to separate out the light components, such as straw, chaff and spikelets, from the heavier grain. The threshing product is dropped gently from a height, and the threshing product container is gently waved in the wind to maximize the operation. It is also maximized by standing on a height (in this case on a specific stand constructed with a triangular piece of wood with three holes into which fit three legs). The heavier material such as the grain falls in a heap directly below the person winnowing, while the lighter material is blown aside where it is often collected to be used as cattle fodder. As winnowing by wind is done, the separation process is helped

Fig. 3-13. Threshing with sticks as a second threshing in Gujarat.

by another person who sweeps the tops of the falling heap, separating out the light smaller seed from the heavier bigger seed. This type of winnowing is a rough separation, and quite often the grain has to be winnowed again by shaking before any further processing can take place. Additionally, often after this stage there is a breakup of the process, and different parts of the products may be sent through different pathways of processing. In general, the byproduct of this stage includes mostly chaff and light seed, both weeds and crops. The product has a higher percentage of grain (both weeds and crop grain), along with heavier plant parts associated with the seeds. The actual winnowing is most often done by men and the sweeping separation assistance is done by women.

Sieving

This processing stage is done mostly in the home bases by women, and quite often in the back or front yard area. In this stage, the *Pennisetum typhoides* and *Sorghum bicolor* seeds pass through the sieve, and the chaff and other larger materials which did not get blown off in the winnowing by wind process are left in the sieve (Figure 3-15).

The sieves are rectangular in size, and their sizes vary dependent on the quantity to be sieved. The diameter of the perforated holes is 7 mm, and the sieve used in this instance was 1.03 m long and 0.63 m wide with a depth of 10 cm (Figure 3-16). The perforated holes are in rough rows, and are approximately 1 cm apart. The sieves are now made of metal, but in many regions of the subcontinent bamboo or reed sieves are still in use. Such devices made with perishable materials would not be preserved archaeologically.

The byproduct from this stage is mostly straw, and heavier vegetative parts of the plant and the seed, while the product is composed of essentially crop-grain, weed seeds, and smaller seed parts. This stage is an efficient method of decreasing the workload in the next winnowing by shaking stage, since the sieving removes the heavier straw that did not separate in winnowing by wind. This separation would be labor intensive in winnowing by shaking, but very efficient in sieving.

Winnowing by shaking (1)

This processing stage occurs in the homes, in the front or back yard areas by women. The pro-

Fig. 3-14. Winnowing by wind.

cess involves the rhythmic tossing, tapping, and shaking of the crop product in a winnowing basket (Figure 3-17). The winnowing basket is deep, box-like and covered with leather. This winnowing basket is markedly different from winnowing baskets of South India, which are wedge shaped (see Figure 3-16). This might affect the efficiency of the process, since the flaying of the basket contributes to the efficient separation during shaking and tapping. The cleaning of the grain is repeated a number of times, and the byproduct, which consists of small crop grains, weed seeds and crop seed parts, is given often immediately to cattle as fodder. The product still has some straw that can be separated only by winnowing with wind.

Winnowing by wind (2)

The third winnowing is similar to the previous winnowing by wind, but the only difference is that the material being processed is of different composition. This winnowing is gentler and slower to avoid excessive loss of grain. The byproduct is mostly straw bits, weed seeds, and small light crop grain.

Winnowing by shaking (2)

This fourth winnowing is similar to the previous winnowing by shaking. It is done at the home base, and is often done at a later date, as a communal activity. The byproduct is fed as fodder to cattle and poultry dependent on the presence and number of these animals.

Pounding

This is done in select cases, when there are a significant number of small crop grains enclosed in the spikelet florets. This is most often done after the second winnowing by shaking 2, and if there are many small grains in the florets, then they are pounded and then processed.

Winnowing by shaking (3)

This stage is related to pounding, and occurs afterwards to separate the chaff dust from the grain.

Fig. 3-15. Sieving *Sorghum bicolor* in Gujarat.

Fig. 3-16. Sieve and winnowing basket used in Gujarat.

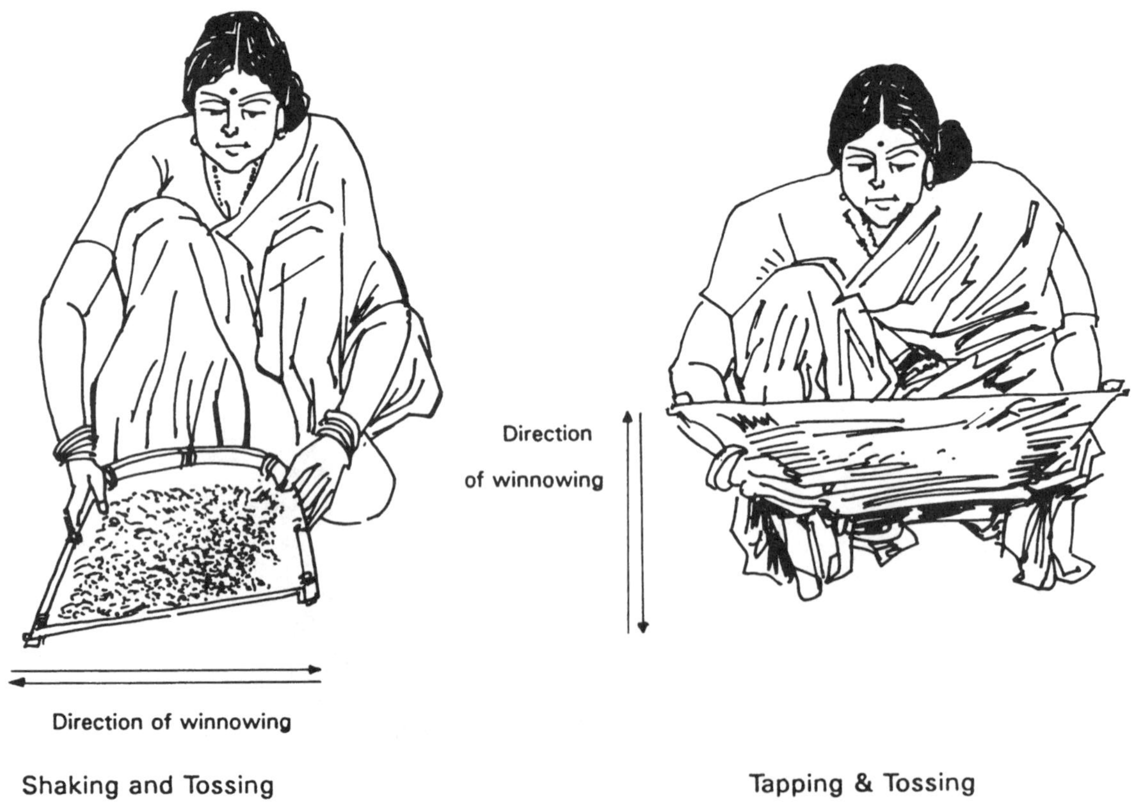

Fig. 3-17. Winnowing by shaking.

The byproduct is chaff dust and it is often small enough just to be swept off the area into the hearth or a trash dump.

Grinding

The grinding stone assemblage is used for this process, which is predominantly done by women, and the objective is to make a flour of the grain for bread and porridge production. The grinding stones are laid out in a rotary grinding system in the homes. The rotary grinding system is placed in a raised wooden stand that catches the flour as it pours out the sides of the rotating grinding stones (Figure 3-18). There is no byproduct from this stage of processing. There is considerable amount of spillage of flour but it is insignificant for archaeological interpretation because it is very unlikely to survive. Such rotary grinding stones are not known from Harappan times, but flat, concave and saddle querns, and various shapes of pestles are quite common.

Use of Pennisetum typhoides *and* Sorghum bicolor

In Gujarat, and at the Babar Kot farm, *Pennisetum typhoides* crop is grown essentially for human food and the byproducts are used for fodder. The byproducts are rarely used for fuel since they do not burn well. When wheat byproducts are not available, then *Pennisetum typhoides* byproducts are used as temper for house and hearth construction. The stalks are more preferable as fodder than the chaff dust. The *Pennisetum typhoides* grain is rarely eaten as seed; it is most commonly eaten as a prepared flour, as a fermented porridge, or as bread.

In some cases, *Sorghum bicolor* crop is grown for two different purposes, and this is dependent on the season of growth. In the Babar Kot area it was noted that *Sorghum bicolor* was grown as fodder in the summer with the use of monsoonal rains. It is also sown in February-March and then irrigated with well water for human consumption.

Fig. 3-18. Grinding stone assemblage used in Gujarat

Typically, more labor and time investment is put into *Pennisetum* cultivation since it has more market value in the region. When *Sorghum bicolor* is grown with the use of irrigation water in February-March then more care is given since more time, money, and energy are invested. More wild fodder is also available for the animals at this time so *Sorghum* is not fed to them. When *Sorghum bicolor* is grown for fodder, it is harvested green and fed to cattle immediately. This period of feeding also coincides with the lean period for wild fodder availability, since most of it is available in the winter months from November through March.

This is an interesting pattern of crop cultivation and is relevant to animal husbandry and seasonality. When there was limited well irrigation in the Babar Kot area, the seasonal river was tapped for irrigation by channels running into the fields that bordered the river. This was done in the recent memory of the farmers and during the previous generation.

Opportunistic Flood Plain Cultivation of Type B Crops *(Panicum miliare)*

Geographical Setting and Economy of the Study Area

The study of 'Opportunistic' cultivation was conducted in the northern part of the state of Andhra Pradesh in South India. Two main study loci were selected for the study and both are set on the banks of the River Pranitha, a tributary of the River Godavari, and are surrounded by fields of crops such as *Sorghum bicolor*, legumes such as *Phaseolus*, and occasionally maize. The selection of the two study areas was based on economies of the dominant social groups, practice of 'Opportunistic' cultivation, and potential for study. Since the area is politically unstable, safety of personnel and timely completion of the study were primary concerns. Therefore, some compromises related to selection of study areas were needed to accomplish the set goals.

The objective of the study of 'Opportunistic' cultivation was to understand the dynamics of this cultivation in terms of the labor, seasonality, crop husbandry, and economic viability of this subsistence system. The two goals of the study were to define the variables that distinguish the different crop processing stages and to understand 'Opportunistic' cultivation as a specific type of subsistence activity to model for Harappan systems.

The agricultural system in the area is predominantly dry farming, being dependent on the seasonal monsoonal rains, rather than irrigation. However, occasionally well irrigation is utilized, a luxury shared by a select few. The area was once densely forested but over the years it has been depleted and reduced to a thorn and shrub 'forest' with minimal heavy foliage. It is therefore conceivable that historically forest products could have contributed significantly to the subsistence base, even though they are almost non-existent today.

The two study locations are quite distinct from each other, even though they lie opposite each other on either side of the river Pranitha. The village of Rapanpalli, lying to the north side of the river, is a very small, nucleated village, while the town of Sironcha is larger and more dispersed. The distance from the village to the river is approximately 2 km, while the town lies along the bank of the river (see Figure 3-1 for details). Agricultural fields surround the village of Rapanpalli, but do not surround the town of Sironcha, since the river runs along one side. Because of this variation in the distance between home bases and fields, logistics vary considerably between the two communities.

Social organization in both the village and the town reflects the classical Hindu tradition, wherein the society is divided into four ritually stratified categories: Brahmana (Priests), Kshatriya (Warriors), Vaisya (Merchants), and Sudra (Laborers). Caste groups are normally endogamous and associated with a particular occupation. The two social groups whose farming practices are studied here, the Harijan and the Kapu, are outside the Hindu ritual hierarchical ideology and are commonly referred to as the 'untouchables'. With modernization, however, there has been a general decrease in the importance of the varna/caste system. Economic differences are becoming more determinant of one's role and status in society rather than ritual differences. In both Rapanpalli and Sironcha, even though the ritual hierarchy is honored, political status and economic class are increasingly more relevant in the social organization. Landowners who are economically independent are quite often the ones who exercise decision making within the community.

The subsistence economy is predominantly farming and fishing, with the latter being an occupation restricted to a particular social group of people from both the town and the village. For this study the location of seasonal fishing is important because it occurs in the same locales as the 'Opportunistic' cultivation, and is an excellent example of several subsistence activities occurring in the same niche. Fishing in this area is affected significantly by the season, and even though fishing could be done throughout the year, there are parts of the year when it is not viable, such as in the midst of the monsoonal rains between June and September. Typically, the fishermen construct temporary huts on the broader western flood plain of the river Pranitha every year as soon as the flood water recedes after the monsoonal flooding (Figure 3-19). The entire family moves into the huts in October and remains there until just before the monsoons begin in June, however the home base is visited occasionally to ensure its safety. The temporary huts on the flood plain are one room structures built of bamboo and reeds. There usually is an outer veranda area where fish are processed for the market; this includes the drying of small fish for both household consumption and distant markets. The fishing is done with nylon nets, since traditional fishing nets are no longer in use. Very often bank loans are taken to buy these expensive nylon nets, and they are often the main investment for the family. Daily catches usually include freshwater fish and shrimp. They are taken to the town market daily and whenever possible taken to nearby towns by bus. Since the duration of their stay on the flood plains is quite long, a significant amount of ashy refuse builds up near the temporary structures of these fisher people. Circular domestic refuse areas are distinctly visible. Nonperishable materials such as pottery, metal tools, fish hooks, net weights, along with domestic activity related discard such as hearth cleanings, charcoal, and fish bones are found in abundance. However, these remnants of the seasonal camps are washed away with the yearly flooding of the river, and would not be visible archaeologically. This is an important observation to note because such groups and their habitation loci could almost never be found archaeologically, but they are significant contributors to the subsistence economy of the society in ethnographic present, and there is no reason to assume otherwise prehistorically.

Farming is the other primary subsistence activity in the region, and it mainly consists of monsoonal rain-dependent dry farming. There are situations however, when a second crop is planted in December utilizing either the remaining ground moisture or well irrigation. Crops including *Sorghum bicolor*, various legumes, and maize are cultivated on the terraces and the higher banks of the rivers. House gardens are very common, usually consisting of a variety of vegetables, legumes and maize, all for household consumption. The household crops are intercropped (the cultivation of more than one crop in a given plot of land) and often the combinations vary according to season and need. These crops rarely enter the market system, and

Fig. 3-19. Fishermen camps on riverbanks of opportunistic cultivation.

are often cultivated by irrigation from the neighborhood water supply, which is most often a well. In the larger land holdings, traditional methods of farming are most prevalent; for example, most land tilling is done with bullocks and a wooden plow. The wealthier landowners use tractors. As with any farming system, there is seasonality in activities, with the sowing and harvesting seasons being highly time consuming, particularly due to the traditional methods of farming and the absence of mechanism. Land tenure as relevant to this paper is unequal, with the lower castes and the 'untouchables' having little or poor land. The Harijan and the Kapu, owners of poor land, are the groups who engage in a specific type of cultivation termed 'Opportunistic' cultivation (Reddy 1991c).

Opportunistic Cultivation

'Opportunistic' cultivation is practiced in the clayey deposits of the active river flood plain. The farming community involved in this specialized cultivation is essentially the economically poorer sections of the each community: the Harijan and the Kapu. They cultivate these deposits as a communal activity with the profit and the labor investment being shared equally. It is practiced to supplement the subsistence, and very often it contributes significantly to the subsistence base, given the low investment of labor in comparison to the cultivation on the banks where land preparation and weeding are heavy time and labor investments.

The crop cultivated in these deposits is *Panicum miliare*, because it seems to be the only crop that can grow in such heavy clay soils. *Sorghum bicolor*, the other millet grown on the upper banks, does not grow in this thick clay deposit, nor is there any other locally known crop that will grow in such deposits except *Panicum miliare*. Typically, *Sorghum bicolor* can only be grown in the sandy clay deposits near the edge of the flood plain, while *Panicum miliare* is grown in the thick clay closest to the water. *Panicum miliare* cultivation is primarily done as an opportunistic activity on the clayey flood deposits on the flood plain (Figure 3-20). Since this is not privately owned, the only initial investment for this cultivation is the sowing time and the cost of the starting seed. According to the

Fig. 3-20. View of opportunistic cultivation along the river banks.

older local informants, there is a long history of such 'Opportunistic' cultivation in the study area. How long it has been an activity specific to the lower castes and the poor landowners is unknown. In addition, there is at least one example from the Gangetic flood plains in North India, where *Panicum miliare* cultivation is done also as an opportunistic activity (personal communication from a local farmer). Such cultivation is significant as it provides insight into a traditional agricultural flood plain activity, which could have implications for understanding agricultural utilization of flood plains in other areas of the world.

The monsoonal flooding of the river occurs during and after monsoonal rains, from July through October. 'Opportunistic' cultivation exploits the summer monsoon rain flooding of the rivers. The river swells and floods the adjacent plain (in a matter of days), and the floodwater often remains for two to three months. When the water recedes (often a slow, gradual process over a month or so), the clay is left behind on the flood plain as a thick deposit that gradually thins from the river to the edge of the bank. That is, the thickest part of the clay deposit is closest to the river and the thinnest deposit is furthest from the river. The thickness of the clay deposits varies considerably from year to year. The deposits stretch along the river course, and their presence is determined by the slope of the flood plain, velocity of the river, amount of flooding every year, and the rate of water recession after the monsoons. At the time of this study, the thickness of the inundation deposit was 1.1–1.2 m closest to the river, and it extended a width of 28 m. The clay is quite well sorted, with almost no sand or silt composition. During the sowing of the *Panicum miliare* crop, the clay is quite viscous, but as it dries up it breaks into well-demarcated blocks. The breadth of the clay deposit is unpredictable and is often dependent on the flooding sequences and the volume of water flooded by the river Pranitha. One of the most important characteristics of these clay deposits is that they are barren of any type of weed seeds, thereby making the deposit in which *Panicum miliare* is sown essentially a sterile clay bed. This is an important feature because weed seeds growing in crop fields very often are included into the crop harvest, and have to be processed out through various processing activities. When and how these weeds get selected out is dependent on their own characteristics and also the processing stage. Weeds have been used (Hillman 1984; Jones 1987) as indicators for the different stages of processing. Therefore their absence has obvious significance in the study of crop processing of *Panicum miliare*.

Botany of Panicum miliare

'Crops' are included within a broad definition of plants cultivated by man for a variety of uses. Not all crops are domesticated. Cultivation means to perform a range of activities related to the caring of a plant and manipulation of the plant for the ultimate use by man. The genus *Panicum* has about five hundred species, but the most important economic species are *Panicum miliaceum* and *Panicum miliare*. Both these species grow well in hot climates and poor soil. They grow rapidly and mature more quickly than any other cereal.

Panicum miliare Lamk. is a hardy plant and capable of producing a crop on the poorest of land and able to withstand conditions of drought and water logging. This is the main reason why it is cultivated on the flood plains in the form of 'Opportunistic' cultivation, because in this form of cultivation waterlogged conditions are typical after sowing. *Panicum miliare* is also called *Panicum sumatrence* Roth ex Roem. & Schult, and it is grown extensively in India but rarely elsewhere. It occurs wild in northern India and southeastern Asia (Purseglove 1979). The husked grain is cooked in the same way as rice and it can also be made into flour. The soft straw is palatable to cattle and the green plant has potential for use as a quick growing fodder. The crop needs little attention and can give a crop, albeit small, even in famine years. It matures in 2.5 to 5 months. It is an annual tufted grass with rather slender culms, 30-90 cm high, soft leaves up to 60 by 2.5 cm, oblong panicles 14-40 cm long, erect hairy branches, spikelets 3 to 4.5 mm long, glabrous, flattened, caryopsis are glabrous, striated brown (Figure 3-21). The plant tillers profusely, and attains a height of about 2-4 feet depending on the variety. The panicles are about a foot in length, with drooping filiform branches. It is a dry-land crop, highly drought resistant, and matures quickly, generally in approximately 80 days. *Panicum miliare* is known as a crop that grows well in poor soils and is often grown as a 'famine' food (Aiyer 1982).

Problem Orientation and Field Methodology

As in the case of the Gujarat project, this study of crop processing stages was designed to define and isolate variables, such as plant parts, weeds and tools, which can distinguish the products and byproducts as distinct assemblages at each processing stage. These can be used for archaeological interpretations, and the definitions of prehistoric archaeobotanical remains. The 'Opportunistic' cultivation system provides a specific system for study otherwise not available. It has potential for comparison with other systems of cultivation, both

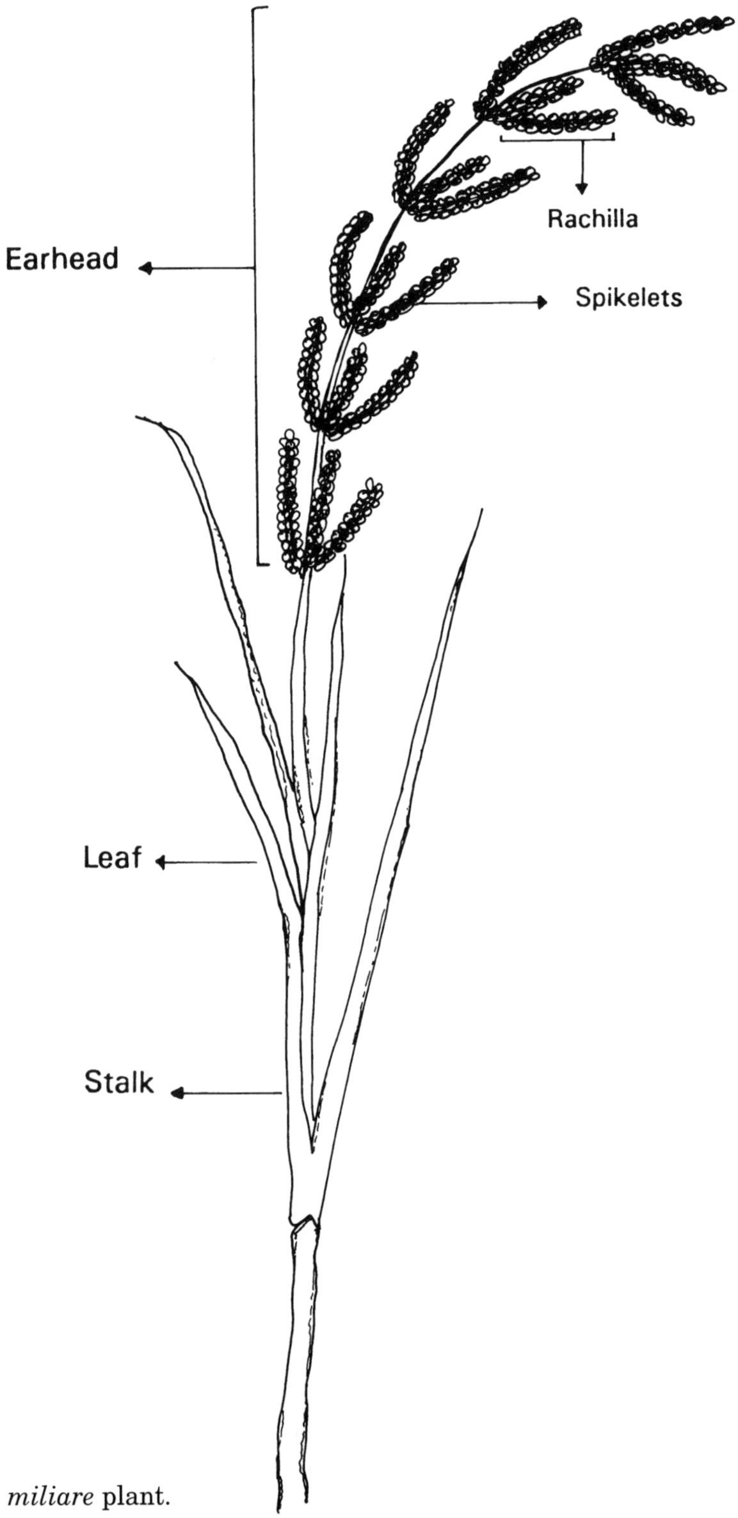

Fig. 3-21. *Panicum miliare* plant.

in South Asia and in other parts of the world. Riverine flood plain cultivation is a very common subsistence activity, and the 'Opportunistic' cultivation provides a possible scenario of subsistence activities on the dynamic river flood plains.

Since the main objective of the study was to examine the crop processing activities for archaeological applications, the crop fields were treated as archaeological sites, and sampling was done within each discrete crop field 'site'. Two stretches of *Panicum miliare* along each side of the river were studied in order to understand the dynamics of this form of cultivation. Two questions were of particular interest: (1) is there any variation within and between the crop fields on each side? (2) does the variability in the width of the flood plain on either side of the river create any observable differences in their fields?

To answer these queries, a thorough reconnaissance was done to identify visual differences within each field and between fields on both banks. The crop fields on both banks were surveyed, noting any particularities such as absence or presence of weeds, density of crop, maturity of crop and other physical features that would be useful parameters of measure. After careful study, it was evident that weeds were present only on the Sironcha bank (Figure 3-22). This was due to the proximity of the field to the terrace bank, which contributed weeds seeds to the field through erosion. All the weeds were low lying, ground cover varieties and were in an earlier stage of their life cycle as compared to the *Panicum miliare* crop. No weeds were noted on the Rapanpalli bank (Figure 3-23).

The fields on both banks showed no variability in the *Panicum miliare* crop distribution but considerable variability in the maturation of the crop. On both the banks, the plants closest to the water were still immature, while those furthest away were ready for harvest. Therefore study units on both the banks were selected based on criteria such as: least trampled, most isolated, closest to bank, or mature/dried.

A total of five study units measuring 5 by 5 m were laid out in the two fields, and then the crop plants were counted in each plot. This count was used to determine significant variables regarding relationships between yield, proximity to water, byproduct compositions, and important differences of the compositions of the plots with known quantifiable differences. To determine the count, the parent plant and corresponding tillers were all counted as one, since they all grew out of a single parent. First, as an experiment to determine an efficient method of counting, the plants were counted before harvest and then after harvest the stubs were counted. Since there was no significant difference, the latter method was employed as it is more time and labor efficient. The five study plots were selected where crops were mature for harvest: closest to river, furthest away from river, and midway (which was also on a path). Selection of these 5 by 5 m study plots was done based on three criteria: least trampled, most isolated, and well dried and matured. Each plot was treated as a separate unit and studied in a similar manner.

Each stage of processing after harvesting was recorded, with specific emphasis on the process description, the quantities of crop product and crop byproduct at the end, the techniques used, and the uses of the byproduct and product when relevant. The weights of the products and byproducts were recorded, and specific quantities of the products and byproducts were taken for later composition analysis. Great effort was made to take homogeneous samples from the product and byproduct piles from each stage of processing. The piles of crop being processed were mixed bottom to top, and then a sample was taken for analysis. This limits any error of bias due to smaller seeds and plant parts trickling to the bottom of the pile, and the larger straw-like material being on the top of the pile. This is a more critical issue in the earlier stages of processing than the latter stages when the seeds were more uniformly weighted.

Sowing

In 'Opportunistic' cultivation *Panicum miliare* is sown in the clay deposits around October when the floodwater has receded, but the clay is still very moist and viscous. If the clay is dry, the seeds will not germinate. The clay deposit has to be a minimum of 0.6 m (60 cm) thick for *Panicum miliare* cultivation. Sowing is done by hand broadcasting; a pot is attached to the sower's belly (for buoyancy and for holding the grain) and seeds are broadcast as the sower wades through the clayey slush deposits. The sowing is done in strips parallel to the river, as the water recedes over a period of time. Because of the staggered sowing, portions of the fields mature at different periods, with the strips furthermost away from the water/river first to be sown and first to mature. This is an optimal strategy for labor investment, since only small portions of the field are ready for harvesting at one time. This planting strategy is also employed in dry farming fields on the banks, plains, and valleys, with some patches sown before others to facilitate timely harvesting and optimization of labor. Once sown, the crop of *Panicum miliare* is not given any other attention in terms of weeding and tending. This is important because it signifies the extensive nature of the practice, with minimal effort and energy invested into cultivation. In the following discussion the different stages of crop processing are presented, as they are relevant to the processing of *Panicum miliare* in 'Opportunistic' cultivation in South India.

Harvesting

Panicum miliare crop is harvested in a Type II method, by cutting a bunch of plants at one time with an iron sickle. Bending down at the waist, a number of plants are gathered in the left hand by the harvester, and then cut with a sickle held in the right hand (in the case of right handed individuals) (Figure 3-24). The stubble left behind on the ground is about 12 cm high. The harvested plants are heaped in a pile and after the harvest-

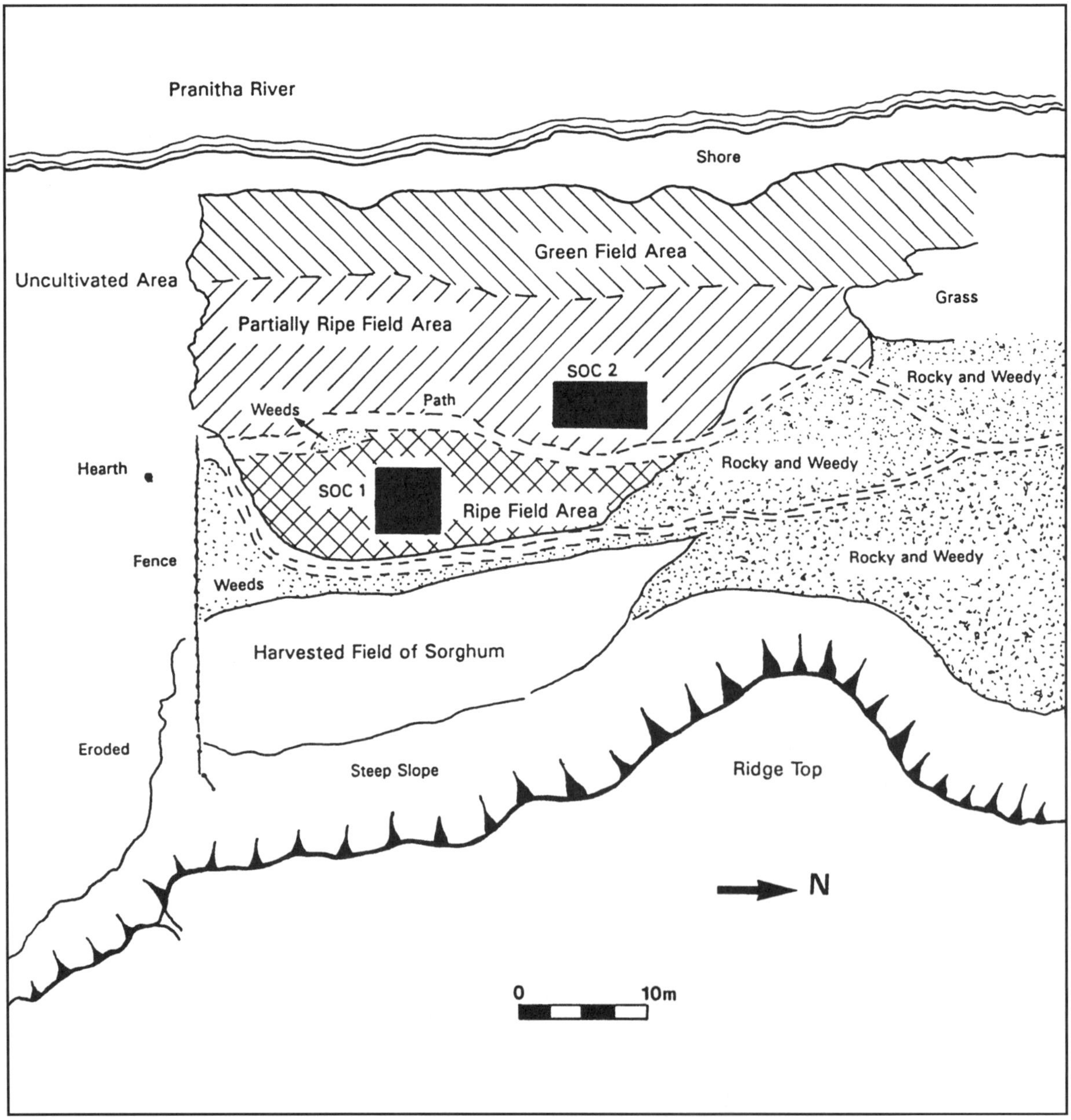

Fig. 3-22. Sironcha 'Opportunistic' cultivation study plots (SOC).

ing is completed, they are then tied into manageable bundles using a stem. Harvesting is done by both men and women.

In the study units on the Sironcha bank, even though weeds were present in the fields, they were not incorporated into the processing products or byproducts, primarily because of the weed types. These prostrate weeds were below the level of the sickle cut for the *Panicum miliare* plants. In cultivation of *Panicum miliare* on the plains, this type of harvesting would result in the inclusion of weeds into the harvested bundles that would not be selected out until later processing stages. This particular 'Opportunistic' cultivation, by the inherent

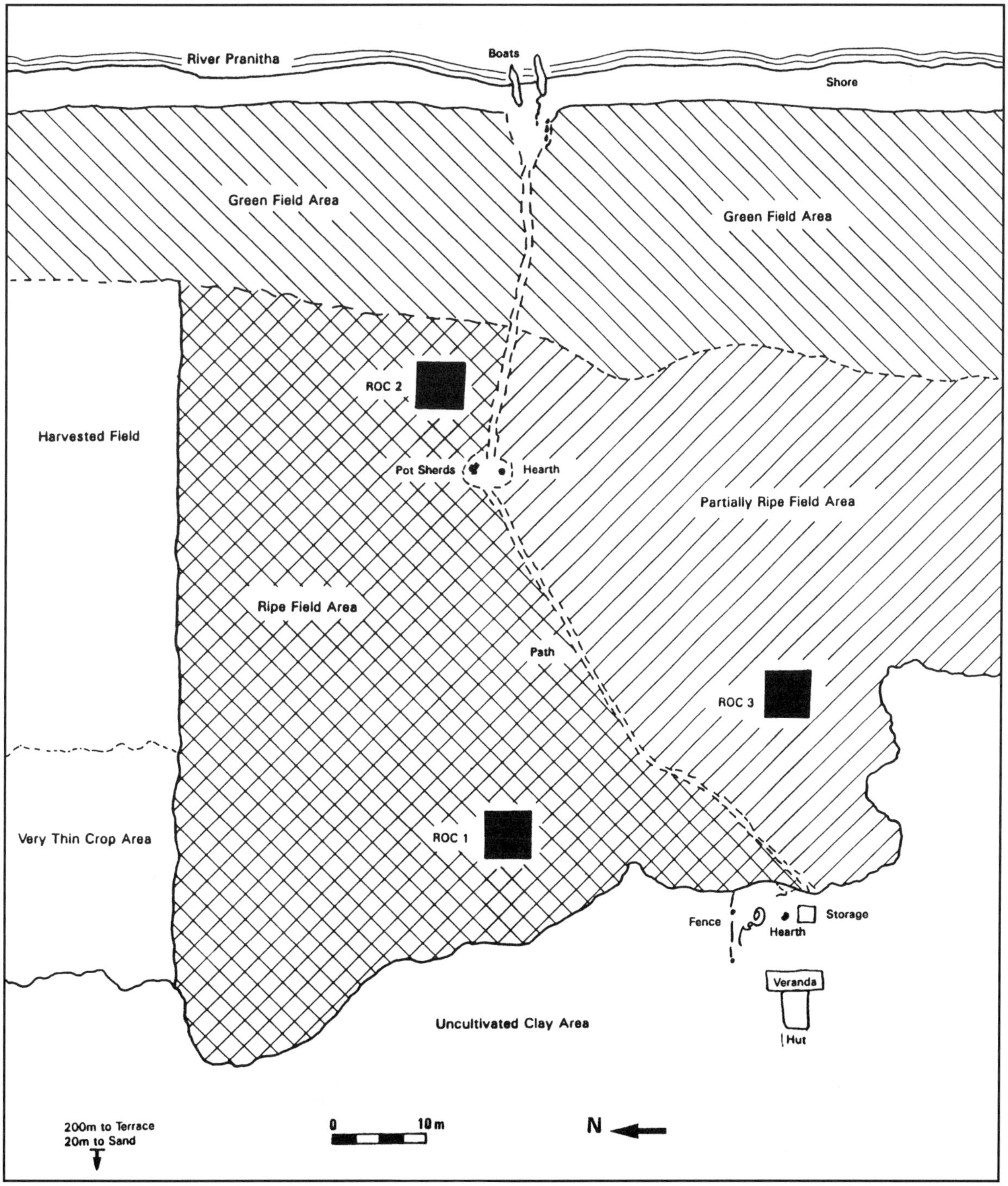

Fig. 3-23. Rappanpalli 'Opportunistic' cultivation study plots (ROC).

nature of the sterile flood deposits, does not present that situation.

This study has also provided new insights into harvest timing and its direct influence on the yield. The preference for *Panicum miliare* is to harvest it slightly green (as it occurred in the study units on the Sironcha bank), while for a crop like *Sorghum bicolor* a more ripe state provides a higher yield. When the *Panicum miliare* crop is well mature and dry, there is a significant loss of grain during the harvest because of the brittle nature of the panicle. The seeds from the panicle head in a green *Sorghum bicolor* crop, however, will not dislodge as effectively as when it is dry, and there will be a greater loss of seeds in the processing. This difference could be an indication of the degree of domestication of each of the crops; *Panicum miliare* being less domesticated as defined by the more brittle nature of the plant's panicles.

The harvested bundles are often laid out to dry in the sun, but this is again dependent on the state of the crop. If it is well dried it need not be dried out before threshing, but if it is partly wet then it is well dried in the sun before threshing.

Crop Processing

Crop processing techniques and the patterning of crop products and crop byproducts after each stage were recorded within each of the five study units. The *Panicum miliare* crop has an average of six main processing steps leading to consumption. These include threshing, sorting, winnowing I (by wind and/or by shaking), pounding, winnowing II (by shaking), and grinding. It was noted in this study, that there are specific tasks for men and women, though there is occasional overlap and sharing of activities. For example, sowing is done predominantly by men, while tending the fields (keeping the animals away) is shared by both. Since the fields are joint cultivation holdings by communities of the Harijan and the Kapu, the activities are not restricted to any single family. Instead, a number of individuals from different families partake in the sowing, tending, and crop processing.

Threshing

Threshing releases the seed along with its attached appendages from the plant. Threshing of *Panicum miliare* is done in one of three different ways: beating with a stick, rubbing and stamping with feet, and/or trampling by cattle. Threshing by beating with a stick and the use of cattle is done predominantly by men, while threshing with feet is most often done by women. Threshing by stick or feet is determined by the maturity level of the crop. If very dry, then threshing with a stick is done, but if the crop is still green then feet threshing is done. The preferred method of harvesting *Panicum miliare* is while it is still green or else much of the grain is lost during harvest. In both methods, the crop plants are laid in a heap on the threshing floor.

The *Panicum miliare* crop in this study was threshed in two different ways: beating with a stick and by rubbing and stamping with the feet. The crop from the Rapanpalli bank, which was drier,

Fig. 3-24. Harvesting *Panicum miliare,* Type II method.

was threshed by beating with a stick (Figure 3-25), while the crop on the Sironcha bank was greener and less dry, so it was rubbed and stamped on by feet (Figure 3-26). The chosen methods of threshing optimize the amount of grain retrieved, the time invested for the amount of crop and, most importantly, the type of crop (in this case *Panicum miliare)*. The seeds of *Panicum miliare* are very small, and using cattle for threshing could lead to considerable processing loss.

Threshing in 'Opportunistic' cultivation is typically done in the fields away from the home base. The threshing floor is often prepared by cleaning the area first, and then plastering it with cow dung. This is often done a few days in advance in order to give it time to dry. The reason a floor is plastered is so that there is minimal loss of grain into the sediment, making it easy to gather up the threshing products through hand sorting. In the subsequent analysis of the threshing products for the two different threshing samples, it is noted that the product of threshing with a stick has more straw and panicles in it, in comparison to the product from threshing with rubbing of the feet. This is significant because it affects the subsequent winnowing process, in terms of amount of byproduct separated out and the time invested in the activity. More importantly, there was also a significant difference in the amount of grain retrieved through the two different threshings, with a greater loss of grain when threshing with a stick.

Sorting

Sorting, the separation of crop product from its byproduct, is done essentially by hand and is equivalent to the raking stages in European crop processing. The byproducts here are the *Panicum miliare* plants that have had the grain removed. They are gathered up from the top of the threshing heap, and piled separately. They are then used as cattle fodder, while the threshing product, which includes the seed, is processed further. Both men and women help in the sorting.

Winnowing I

After threshing, the crop is winnowed to separate the grain from chaff, straw, and spikelets. *Panicum miliare* is winnowed by shaking with the use of a basket (Figure 3-27). This method is specifically chosen because of the small size of the crop seeds. Winnowing by shaking is more appropriate for these small sized seeds because it can be controlled effectively (as compared to winnowing by wind) to avoid excessive loss of the crop grains. This activity is done in domestic contexts, and very rarely on the threshing floors. The straw, chaff, and spikelets that are separated from the grain through winnowing are then added to the cattle fodder. The winnowing process stage involves a series of winnowing (regardless of method of winnowing). The series of winnowings cannot be re-

Fig. 3-25. Threshing *Panicum miliare* with sticks.

garded as separate winnowings because they are done sequentially (without them being considered as separate stages). The series of winnowings continue till the product is cleaned of all the non-desirable components.

Winnowing by shaking is done using a winnowing basket, which is woven, and rectangular wedge-shaped. It is thin and smooth, and often plastered with dung and *turmeric* (a disinfectant spice) to ensure a smooth surface. The grain is placed in it, and then by the rhythmic tossing of the grain, followed by the shaking of the winnowing basket, they are separated from contaminants such as straw, chaff, spikelets, or weeds. The cleaning of grain is repeated many times until there are no contaminants. This is a time-consuming and labor-intensive process, but there is very little loss of grain (unlike winnowing by wind) and the process is very effective for separating grain from chaff. This activity is done predominantly by women. If the grain is not used immediately, it is stored or taken to the market after this stage of processing.

Fig. 3-26. Threshing *Panicum miliare* with feet.

Pounding

After the grain has been cleaned, it is pounded to separate the seeds from the husks (Figure 3-28). This activity always occurs in domestic contexts - inside or outside the house. The grain is placed in the depression of a grinding stone and then pounded with a thick heavy wooden staff. Through repeated pounding, the seeds are separated from the husks. If the grain is partially ripe when harvested, quite often it is roasted over a fire before pounding to facilitate the efficient separation of the husks. If the grain is well ripened at the time of harvest, it is dried in the sun before pounding. This distinction is important for archaeological inferences since it has the highest probability of accidental burning and preservation in archaeological record. Often there is much spillage of grain during this processing (Figure 3-29). Minimal effort is made to clean up the spillage if the processing is done outside the house, however, when done within the house, all spillage is cleaned up and incorporated into the product.

Winnowing II by shaking

Once the seeds are freed from the husks through pounding, winnowing by shaking is done to separate out the seeds. The fine chaff dust that is separated is an attractive cattle fodder. Often the seeds are cooked similar to rice after this stage of processing.

Grinding

In some instances, the seeds are further pounded into a flour and then consumed as a porridge-meal. This is done through a process similar to pounding. One of the differences between the two processes is that there are no byproducts from grinding, for even though there is spillage, it is not significant in terms of quantity and archaeological preservation. *Panicum miliare* flour can also be made using a rotary grinding method. This entails two circular stone slabs set one upon the other. The grain is dropped through a hole in the center of the upper stone and while the upper stone is rotated by hand, and ground flour then flows out the sides.

Use of Panicum miliare *byproducts*

All the *Panicum miliare* byproducts from threshing, winnowing I, and winnowing II (including straw, panicles, chaff, and chaff dust) are used as fodder. This material is not storable for long periods, and the forage has its highest nutritive value when fed fresh soon after the crop processing. *Panicum miliare* byproducts are never used as fuel. The comparative low nutritive value of *Panicum miliare* grain and its non-storability make it a less attractive crop to grow primarily for human consumption. Instead it is ideal as a short-term human food supplement when grown in an opportunistic manner.

Discussion

'Opportunistic' cultivation presented here stands as an important form of traditional agriculture. It has historical longevity in the area, particularity in terms of the clay deposit, type of crop cultivated, nature of cultivation, and the absence of weeds in the fields. All of these factors contribute to making it an important case for the further understanding of flood plain agricultural systems (past and present), communal labor, and investment. Specifically, new insights have been gained regarding existent flood plain agricultural systems. The communities involved in this cultivation belong to the lower castes, who also happen to own the poor non-irrigated land. Only this sector of the rural society partakes in 'Opportunistic' cultivation; and their participation in this cultivation is uncertain if their land holdings were irrigated and high yields of *Sorghum bicolor* were possible. One could suggest a decline in 'Opportunistic' cultivation given this scenario. However, this cultivation might persist just as the name 'Opportunistic' suggests, and perhaps just as a crop for fodder.

The implications of 'Opportunistic' cultivation for ancient flood plain or riverine agricultural systems are more significant. Given the number of riverine and flood plain agriculture instances in the world, this example of a specific type of cultivation of *Panicum miliare* potentially could be relevant to other situations. Of particular interest to the study of ancient agriculture is the absence of weeds in the flood plain deposits and in the subsequent crop fields. This affects the entire range of crop processing activities and the resulting patterning in the products and byproducts from each stage of processing. Ethnographic crop processing studies have been used to make inferences on the

Fig. 3-27. Winnowing by shaking of *Panicum miliare*.

archaeological use of specific crops (Hillman 1981; Jones 1987; Reddy 1991a, 1991c). Such studies are based on the observations that each step of crop husbandry and grain processing has a measurable effect on the composition of crop products and byproducts. These effects are studied and 'cause-and-effect' models built for archaeological interpretation. It is often the case that weeds are indicators of different stages of processing (Jones 1987). However weeds are generally absent in this study of 'Opportunistic' cultivation, due to the sterile nature of the clay flood deposits. Even when weeds are present, they are not incorporated into the crop processing activities. Furthermore, this type of 'Opportunistic' cultivation is particularly relevant for reconstructing ancient agricultural systems of the Nile, the Ganges, and the Indus rivers, all of which have seasonal flooding, and a long history of riverine and flood plain agriculture.

Specifically in a riverine system like the Indus, where there were major urban centers and smaller 'rural' settlements during the Harappan period (2500 - 2000 B.C.), such opportunistic, extensive cultivation is more likely to occur in the latter type of settings. Such cultivation, although practiced sporadically in the urban environs (probably by

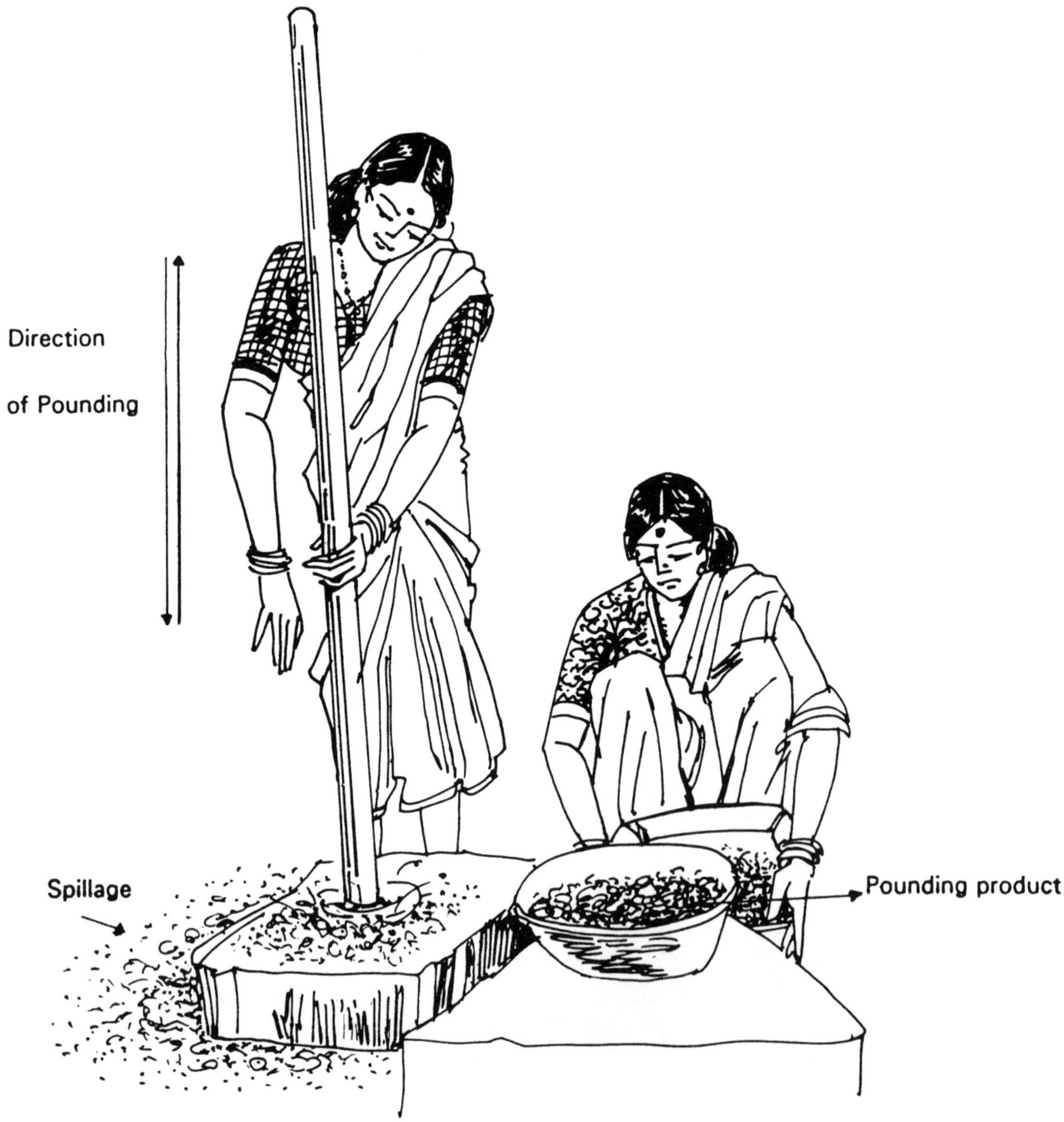

Fig. 3-28. Pounding of *Panicum miliare*.

lower status groups), would have contributed minimally to the overall subsistence economy. The low yields per acre and the heavy time investment in crop processing would make it less attractive in the urban centralized market economy where intensive agricultural systems were more successful. 'Opportunistic' cultivation on the flood plains would be an attractive option in the rural settlements of the Harappan period, where there was a gradation between intensive and extensive agricultural systems. The degree of importance of animal husbandry would also be a contributing factor for the selection of such cultivation. In such cases, settlements optimally located near a flood plain could very well exploit the local flood plain opportunistically, thereby supplementing their subsistence resources for human consumption or animal fodder. Such cultivation would be particularly attractive for pastoralists who often grow quick maturing millet crops both for themselves and their animals.

'Opportunistic' cultivation though observed specifically in one river flood plain in South India, has important ramifications for studies related to ancient agriculture, modern rural development, and communal economic activity. With modernization, many traditional practices of crop husbandry discussed here are slowly being replaced by the mechanized agriculture of the 20th century, thus making studies of crop processing activities increasingly important for the further understanding of ancient agricultural and crop processing practices.

Crop Processing Stages and Archaeological Interpretation

Even though the processing pathways are quantifiable in terms of compositions and specificity of distinctive variables, it is important to elucidate the likelihood of these stages being identified and discovered archaeologically. There are four aspects of each processing stage which need to be considered to define the archaeological relevance: storage, the probability of accidental burning at or before the processing stage, occurrence in domestic contexts, and the likelihood of preservation and survival into the archaeological record. Table 3-1 summarizes this classification.

Storage is an important variable, because storage of a crop product or byproduct (at any stage) increases its chances of accidental burning and survival into the archaeological record. Additionally, since storage is most often in domestic contexts the chances of discovery are increased. It is important to note that the storage index of a crop product or a byproduct is variable and constantly changing, but the general pattern remains. For example, the likelihood of a threshing product being stored is generally low, however, it is possible that occasionally storage index of this assemblage could be moderate but it is never high.

Survival into the archaeological record is based on whether there is carbonization of the different components of an assemblage through burning. Therefore, the probability of accidental burning

Fig. 3-29. An example of pounding spillover (*Sorghum bicolor*).

occurring at the processing stage is very important, since this index determines which processing stages are by nature more discoverable in the archaeological record. The process that determines this index is roasting which occurs after sieving. The survival indices of the subsequent stages' products and byproducts are therefore high.

Another factor that determines the carbonization rate of an assemblage is the likelihood of the processing occurring in domestic contexts. Domestic contexts are locales of intentional fires (cooking, heating, etc.) that increase the probability of accidental burning. Additionally, when a stage occurs in domestic contexts, there is a higher chance of the byproducts being included into contexts of burning or charring. Domestic areas are in general discovered more often archaeologically as compared to crop fields.

The crop products and byproducts of different processing stages are thus ranked (unlikely, low, moderate, and high) to determine which assemblages are archaeologically significant. Threshing product occurs away from domestic contexts and has a low storage rank. Accidental burning occurring at this stage is unlikely and the assemblage is unlikely to be preserved and discovered archaeologically. The threshing byproduct has similar ranking and is not significant for archaeological contexts.

Winnowing I (by wind) product has a moderate storage likelihood, but because accidental burning is unlikely, and the low chances of the process occurring in domestic contexts, the survival and discovery of this assemblage are relatively low. Winnowing by wind byproduct is not stored and therefore unlikely to be prone to accidental burning; therefore, its survival and discovery in the archaeological record are unlikely.

Storage of the sieving product is highly likely, but it is not prone to accidental burning through the processing. However, it occurs most often in domestic contexts, so its archaeological discovery and survival probability is moderate. Storage of sieving byproduct is low, and it is not likely to be burned during the process. Yet since the process occurs most in home bases, the probability of archaeological discovery and survival is high.

The winnowing by shaking products and byproducts are archaeologically significant, mainly because of the location of processing in domestic contexts. The winnowing by shaking product has a high storage ranking and high probability of accidental burning (because it is roasted before pounding), and thus the likelihood of its archaeological survival and discovery is high.

The pounding stage is preceded by roasting and the pounding product also has a very high accidental burning likelihood, and its archaeologi-

Table 3-1. Crop Processing Stages and Archaeological Relevance.

Crop Processing Stage Product and Byproduct	Likelihood of Assemblage Storage	Probability of Accidental Burning	Occurrence of processing in Domestic Contexts	Probability of Preservation and Discovery
Threshing Product	low	unlikely	unlikely	unlikely
Threshing Byproduct	low	unlikely	unlikely	unlikely
Winnowing I Product	moderate	unlikely	low	low
Winnowing I Byproduct	unlikely	unlikely	low	unlikely
Sieving Product	high	unlikely	high	moderate
Sieving Byproduct	low	unlikely	high	moderate
Winnowing (s) Product	high	high	high	high
Winnowing (s) Byproduct	unlikely	moderate	high	high
Pounding Product	unlikely	high	high	high
Winnowing II Product	high	high	high	high
Winnowing II Byproduct	unlikely	high	high	high
Grinding Product	high	high	high	unlikely
Grinding Spillage	unlikely	high	high	high

cal survival and discovery are high mainly because of its domestic contexts despite of its low storage probability. Winnowing II (by shaking) product has very high ranking for all the four variables, and this makes it a very significant assemblage for archaeological interpretations. Grinding product has high storage and high accidental burning ranks, and it occurs in domestic contexts. Archaeological survival and discovery are unlikely, however, since the product of grinding is flour, which does not survive into the archaeological record. Thus, it is apparent that the archaeological significance of the different stages is extremely varied, and these differences need to be considered when making archaeological interpretations.

Summary

Ethnographic studies of crop processing in India (Gujarat and Andhra Pradesh) have demonstrated that there is a significant difference in processing pathways of the crops examined. For example, sieving is very prominent in the processing of Type A crops but absent in the 'Opportunistic cultivation' of Type B crops. Pounding is an important stage in the Type B processing but is of minimal importance for Type A crops. This variation is quantifiable since it creates distinguishable compositions within the byproducts and products. These differences in processing are a result of several factors including regional variation, type of cultivation, crop type, use of product, and seed morphology.

Regional variation is important to consider, especially for archaeological interpretations. Even though Hillman (1984) and Jones (1987) argue for a limited number of ways for processing a plant, it is clear from this research that there are other methods (such as winnowing by shaking, pounding, and grinding) that differ from previous observations made in traditional societies of Turkey and Greece. This regional variation in methods could be the result of historic developments. For example, Gujarat is closer to other culturally distinct regions of Pakistan and Afghanistan while Andhra Pradesh is more of a cultural island in peninsular India, and it has had less influence from outside sources. These differences could also be an expression of ethnically distinct groups who process materials a certain way to maintain group identity and ethnicity. These factors should be kept in mind when building models for archaeological interpretations.

I argue, however, that crop seed morphology plays the primary role in determining which processing method is chosen. This is particularly relevant in the case of winnowing. As discussed earlier in the chapter, winnowing can be done in two ways: by wind or by shaking (in an open wedge shaped basket). The exclusive choice of winnowing by shaking method for *Panicum miliare* was based on the small size of the crop seeds, which would be lost through winnowing by wind. However, in the case of *Sorghum bicolor* and *Pennisetum typhoides* winnowing by wind was chosen as the first winnowing process (later followed by winnowing by shaking) because the crop seeds were not small and would not be lost during the process. Therefore, it can be asserted that for archaeological applications, the winnowing choice is determined by the crop grain in question.

It is important to note that winnowing by shaking can be compared to Australian yandying as described by Cane (1989). Australian yandying involves a large wooden dish that is rocked and shaken in various directions at different angles to separate chaff and other rubbish. In general, yandying and winnowing by shaking are similar processes since both involve intensive processing through shaking, rocking and tapping, and produce similar compositions in byproducts and products. Cane (1989:105) suggests that yandying replaces winnowing and sieving of western Eurasian grain processing. Similarly, in the *Panicum miliare* processing pathway, winnowing by shaking replaces the winnowing by wind and sieving processes in the processing pathways of *Sorghum bicolor* and *Pennisetum typhoides*. This choice of winnowing method is a result of the crop seed morphology. In both the studies (opportunistic cultivation and summer cultivation), winnowing by shaking replaces the fine sieving stage present in the Near Eastern regions as described by Hillman (1984) and Jones (1987). This second situation, where winnowing by shaking replaces fine sieving, is a result of both ethnic choice and crop seed morphology.

The type of cultivation and the type of crop grown are also deciding factors in the processing. This research in India has clearly demonstrated this distinction. In 'Opportunistic' cultivation, the nature of the cultivation itself determines whether weeds are included and therefore the amount of processing done is minimized. In contrast, the inclusion of weeds is unavoidable in the Gujarat monsoonal cultivation, and the processing has to take this into account. The type of crop being cultivated also affects the processing pathways. There are a variety of harvesting choices in Type A crops, such as *Pennisetum typhoides* and *Sorghum bi-*

color, as discussed earlier. The choice of harvesting affects the processing pathways regarding the inclusion of weeds, the amount of byproduct cleaning, and the number of cleanings done at each stage. In Type B crops, such as *Panicum miliare*, there is no choice in harvesting method. The seed morphology, whether it is a hulled seed crop as in *Panicum miliare* or a 'naked' seed crop as in *Pennisetum typhoides* and *Sorghum bicolor*, is a deciding factor for whether certain processing stages are needed or not. This is evident in the stages presented for each of these crops. For example, the dehusking process by pounding is not needed for *Pennisetum typhoides* and *Sorghum bicolor*, but it is an essential processing stage before consumption of *Panicum miliare*.

4. The Search for Patterns: Ethnographic Modeling and Archaeological Relevance

The ethnographic research aims at isolating the distinguishing characteristics of crop processing activities and their archaeological implications. This chapter presents the composition analysis of the ethnographic materials collected from processing stages of several millet crops. It also presents ethnoarchaeological modeling for two different types of crops studied: Type A and Type B (Table 4-1). Finally, variations in processing are modeled with respect to archaeological implications when crops are cultivated for three different purposes, for food only, for food and fodder, or only for fodder.

Kramer (1979:1-4) has explained that observations of contemporary behavior can be utilized to develop and refine insights into past behaviors, and this is particularly relevant when strong similarities can be shown to exist between the environments and technologies of the past and contemporary sociocultural systems that are being compared. This is especially true when one is considering the behaviors related to crop processing activities where the nature of processing is as much a factor of cultural tradition as it is of the morphological structure of the crop (Reddy 1991c). Ethnoarchaeology examines contemporary sociocultural behavior from an archaeological perspective with ethnoarchaeologists attempting to systematically define relationships between behavior and material culture patterning. These behaviors and observations are not often explored by ethnologists, and what is distinct about ethnoarchaeologists making these observations is that they ascertain how certain features of observable behavior may be reflected in the remains that archaeologists may find (Kramer 1979:1-4). Once such observations have been made, the ethnographic data can be used in a number of ways. Stiles (1977) suggests three general uses for such data: ethnographic analogy, the generation of hypotheses or models, and lastly the testing of hypotheses. This project involves each of these uses and this chapter deals with the first two methods through ethnographic modeling.

Ethnographic modeling entails modeling of ethnographic data for archaeological interpretations through rigorous correlation studies and other data analysis. It involves the generation of hypothesis and models, with ethnographic data being used as a source of information to generate these hypotheses and models about the past. Hypotheses are generally postulated in response to the use of ethnographic information as a potential analogy to a specific archaeologically observed phenomenon, but they need not be (Jones 1983). Models, on the other hand, seek to represent or describe a set of observed phenomena from the real world. A body of ethnographic observations may be combined to aid reconstruction and interpretation of a corpus of archeological data. From analogies made between the archaeological material of a particular site or 'culture' and the ethnographic parallels of a model, a selected segment of prehistoric life can be reconstructed.

The testing of hypotheses is essential, and this is done after the hypotheses have been formulated

Table 4-1. Harvest Methods and Crop Types.

		HARVEST METHOD **I** (cutting at top)	 **II** (cutting at base)
CROP	**A** (Summer Cultivation)	*Pennisetum typhoides* *Sorghum bicolor*	*Pennisetum typhoides* *Sorghum bicolor*
TYPE	**B** ('Opportunistic' Cultivation)	NONE	*Panicum miliare* *Setaria tomentosa*

from ethnographic observation. If the pattern of the archaeological material falls within the range of variation of the pattern of the physical traces that resulted from the ethnographic activity, one can then assume a certain probability that a valid analogy has been made. This degree of probability cannot be measured precisely, but the strong correlation of the two conditions should be indicative of an appropriate and viable analogy.

Thus, ethnographic materials were collected from all of the crop processing stages that were archaeologically relevant. Therefore, samples of pounding byproduct were not collected because the dominant components of these assemblages, such as chaff dust, are not identifiable macroscopically, and are unlikely to survive into the archaeological record. Similarly, threshing byproduct was not collected since its components, straw and stalks, would rarely survive into the archaeological record as a distinct activity assemblage since they are given as fodder to cattle. Thus, only samples from archaeologically pertinent stages were analyzed for their compositions, and subsequently using these results, ethnoarchaeological models were built for the processing pathways of each crop.

In this chapter the general methodology and data analysis procedures will be presented first, and this will include a discussion on weed categories and crop types. Subsequently, the data analysis of Type A crops (summer cultivation of *Sorghum* and *Pennisetum*), and a model for their cultivation is discussed, followed by the data analysis of a Type B crop cultivation without weeds ('Opportunistic' Cultivation of *Panicum*) and a model for its processing. Finally, data analysis and a model for the processing of wild plants such as *Setaria tomentosa* is presented. The chapter concludes with a summary, and discussion of three systems of cultivation: crops cultivated as food for humans, fodder for animals, or as both.

Methodology and Data Analysis

Harvesting of the study plots was done by farmers and farmer's aides who were familiar with the processing activities. As discussed in the previous chapter, a range of study plots was selected, with each study plot being treated as a separate unit and studied in a similar manner. This facilitated computations regarding relationships between yield, proximity to water, byproduct compositions, and important differences of the compositions of the plots with known quantifiable differences.

The underlying premise for the ethnographic sample analysis was to identify and isolate distinguishing variables that are specific and distinctive to a particular stage or process. Therefore the main objective of the crop products and byproducts sample analysis is to establish the composition of each of these samples in relation to specific stage processing activities. All samples analyzed for composition were treated in a similar method. The major proportion of this initial analysis includes the basic documentation of assemblage composition from each processing stage. The samples were thus sorted and quantified for their compositions with specific attention being given to the range and quantity of weed species. To build models for archaeological applications, the identification of weed seed characteristics relevant to the different stages of crop processing are necessary. Therefore, during analysis of the samples from each stage product and byproduct, special emphasis was placed on recording weed sizes, their tendency to remain in heads, their weights, and their frequencies. Specificity of occurrence at a particular stage was recognized as an important variable, and an attempt was made to identify this during the data collection. Since weeds and crop plant parts are the main criteria for identification of crop processing stages in this study, careful attention was given to the selection of the different components for the data collection and composition analysis.

Materials collected during both the seasons of ethnographic fieldwork ('Opportunistic' cultivation in Andhra Pradesh and summer monsoonal cultivation in Gujarat) were analyzed to enable broad generalizations and comprehensive conclusions about crop processing activities and patterning. Consistency was maintained throughout the study of ethnographic crop processing samples.

The first stage in the analysis of each processing sample was sieving through three geological sieves: Sieve A (4 mm), Sieve B (2 mm) and Sieve C (0.425 mm). This was done to facilitate the identification of the variation in the physical sizes of the different components (such as rachilla without spikelets, rachilla with spikelets, specific weed species, and so on) for each stage's product and byproduct. Instead of measuring the size of each different component, distribution within the sieve sizes was used. First, the volume of the sample was recorded, then the three sieves were then placed on top of each other (with Sieve A on the top and sieve C at the bottom), and the sample was poured onto the top sieve (Sieve A) in lots of 1 liter to minimize error through spillage and effective sieving. Once the sample was sieved, the three sieves' contents were collected separately in labeled bags and set aside for composition analysis. For example, 64 samples, collected in the field from *Pennisetum typhoides* processing, yielded 192 sub-samples (from the three sieves) for analysis.

In the next step of analysis, measurable amounts of the sized sub-samples were sorted into the distinguishable botanical categories. Each sub-sample was poured onto an open plate, and then the different components were sorted out and collected separately. After the sorting of a subsample was completed, the different components were quantified (actual counts, weights, and volumes). After quantification, they were bagged, labeled, and stored for future reference if needed. When relevant, the number of individuals in each category was counted. General observations of patterns and distributions were recorded during the process of sorting and measuring. Once the measurement and recording of the data were completed, the studied sub-samples were bagged, keeping the categories separate, and stored. The data (actual counts, weights, and volumes) were then analyzed through quantitative and qualitative pattern searching.

In this study, the crop processing sequences are broadly divided into "higher processing stages" and "lower processing stages". Lower processing stages refer to the initial stages that occur in the fields and home bases. Additionally, in a non-mechanized and traditional economic system when cultivation is geared to both household consumption and trade or exchange, the marketing of the crop products is done immediately after the lower processing stages. Higher processing stages refer to those stages which are closer to the consumption product or final product, and occur primarily in home bases or domestic activity areas. Importantly, when a crop product is intended for marketing (through trade or exchange) it is very rarely put through the higher processing stages. For example, the lower processing stages of a Type B crop (such as *Panicum miliare, Setaria italica*) include threshing, sieving, and winnowing I. Both sieving and winnowing I are most often done in the home bases while threshing is done in the fields. Trade or exchange of Type B crop products occurs after winnowing I in most instances. Higher processing stages of Type B crop include roasting, pounding, winnowing II, and grinding. Often the later series of winnowing I are repeated just prior to roasting and pounding. The distinction between the lower and higher processing stages is significant for ethnoarchaeological modeling in relationship to likelihood of archeological inclusion, survival and discovery, and very importantly the identification of cultivation versus trade and exchange.

The importance and significance of this methodology lie in its replicability and testability. Many of the components could at a later date be collapsed, but for this initial stage of composition analysis, all of the recognized components are the basic elements of each sample. In the subsequent discussion, emphasis will be placed on the archaeological relevance of each component.

Weed Categorization

Samples of weeds were collected from the study fields during the summer monsoonal cultivation crop processing study in Gujarat. It was necessary to collect these weeds in order to have a comparative collection for the analysis of the harvested samples of the ethnographic products and byproducts. The weeds associated with the millet fields are particularly important since they could be filtered out by different processing stages and methods, and thus could be used as indicators of specific products and byproducts at different stages of crop processing. The weed collection was done in a systematic manner through initial reconnaissance of the fields followed by interviews with the farmer on weeds and their associations with the different crops. During the collection of the different weeds in the fields, notes were taken regarding their location (in terms of their association with a specific crop field), local common names used by the farmer, and their general description. None of the weeds collected were from within the study plots, so as not to affect the frequency of weed plants being included into the crop processing stages. The plots, however, were carefully scanned visually to make sure that all varieties of weeds had been included into the comparative collection. The taxonomic identification of the weed samples was done through consultation with botanists (specially Ms. Charul Joshi) at the Laboratory of Environmental Sciences, Department of Biosciences, Saurashtra University, who were familiar with the local plant geography.

Previous ethnographic studies conducted by Hillman (1984) and Jones (1987), particularly designed to model for archaeological interpretations, used weeds as indicators of different crop processing stages. Jones (1987) used weeds more exclusively than Hillman (1984, 1985). Through their studies of present day crop products, they argued that the principal factors determining what seeds were present in any one crop product were: (i) the ratio of their surface area to weight (their winnowability), (ii) seed size (sievability), and (iii) seed 'headedness'. Thus each weed species present in the crop processing stages was characterized in terms of these three variables.

Hillman (1984, 1985) stated that winnowability, the probability that a seed can be winnowed out of the prime products, seems to depend primarily on the ratio of its surface area to weight. Therefore, the bigger and heavier a seed the lower the probability of being winnowed out (i.e., a lower winnowability). The presence of dispersal appendages such as hairs or wings increases a seed's winnowability.

Sievability is dependent on seed size, and there is a clear correlation between sievability and winnowability, as also determined by seed size/weight ratio. Hillman (1985) groups seeds and fruits into four sievability/winnowability classes, each of which is characteristic of a single class of crop product. Therefore, he argues that in the absence of mixing, the identification of these particular products is very straightforward. The classification of weeds into the four classes is based on weed sizes. These are not absolute sizes, but relative to the width of the primary spikelets of the wheat crop. This classification is not relevant to millet processing, since most of the millet crops being considered in this study are of different size and shape when compared to wheat.

Headedness is the measurement of a weed seed being contained on a head (or panicle), and often the weed seed is released from the head only after a certain stage of processing. This is particularly true if the weeds are immature at the time of harvesting. Hillman and Jones both argue that theoretically the two parameters 'winnowability' and 'headedness' should not be used independently.

Jones (1983) employs a similar weed classification with the size of the weed seed being relevant for fine sieving processes; the tendency of weed seeds to remain in heads, spikes or clusters despite threshing, being relevant to coarse sieving, and the aerodynamics/winnowability being most relevant to winnowing.

The weed classification of Hillman and Jones had to be modified since the crops being studied are the millets, *Sorghum, Pennisetum, Panicum,* and *Setaria*. Modification was necessary not only because of the different crops being studied, (which affected the weed size categories made by Hillman and Jones), but also because the processing stages and methods were significantly different. For example, Hillman's crop processing stages included a fine sieving stage, which was absent in the crop processing pathways studied in India. Sievability was an important factor for Hillman to consider, but in India it is not of such critical importance. Similarly, an additional stage (winnowing by shaking) was observed in India but was absent in Hillman's crop processing studies. In this study a winnowing basket separates crop products and byproducts through shaking, tossing, tapping, and hand sorting; therefore, seed headedness, seed weight, and seed size are all of equal importance in this winnowing stage.

The weed classification developed in this study incorporates some of Hillman and Jones's criteria for the selection of factors affecting the presence of seeds in the crop products. In addition, criteria that were more relevant for millet crop processing (with respect to both the crop and the processing methods) were used to develop a classification. This classification does not use terminology that is implicative of a stage (as suggested by the classification terms of Hillman and Jones—winnowability, sievability). Instead it uses attributes of the weeds that can have implications at various stages depending on the processing pathway and what the processor is selecting for specifically. Therefore, this classification is not dependent on stages, but rather concentrates on the attributes of the weeds and the processing method. For example, whether a seed is winnowable or not is irrelevant, because winnowability (through shaking) is not dependent on one characteristic, but a combination of characteristics. One of the main advantages of this classification is that it does not use weed size relative to crop seed dimensions; therefore, the classification also can be used in the future for other crop types.

During the analysis, all the weeds collected were classified into size, seed headedness, and weight/aerodynamic categories, which were then used to classify the weeds into distinct categories. Each category, its significance in the crop processing activities, and the method of categorizing will be considered in the following discussion.

Size

Size is an important variable because it is often the first discriminating factor used to select against non-crop seeds (i.e., weeds). The bigger the seed the higher the probability that it will not be present in the latter stages of processing. However, size is not an independent variable, since the combination of size and weight is often the determining factor. It is nonetheless important to note that winnowing by wind tends to eliminate very small seeds whose weight is not a controlling factor. Seed size also affects sievability as argued by Hillman (1984). However, in India sieving is very coarse and aimed more at the separation of very large non-crop seed components, such as unthreshed/immature inflorescence heads, large spikes, and stalks. This coarse sieving does not separate out weed seeds of any size efficiently. Seed size is a determining factor in winnowing by shaking but unlike winnowing by wind there is more control on selection. Even small seeds can be selected for retention in winnowing by shaking, just by altering tapping, shaking, and hand sorting movements and eliminating tossing. Therefore, seed size is not the sole discriminating factor in winnowing by shaking.

In this study, the weed seed size was determined through the use of geological sieves. As discussed earlier in this chapter all samples to be analyzed were first sieved through three geological sieves: Sieve A (4 mm), Sieve B (2 mm) and Sieve C (0.425 mm). This was done to facilitate the identification of size variations in different processing stages. Weed size categories were then formulated using these three sieve sizes and the ethnographic sieve used in the crop processing. The size categories range from very small to very big, and coded from 1 for very small to 5 for very big. Each weed species was thus categorized into the appropriate size category depending on which sieve it was primarily associated with. Size categories 1 and 2 were classified as small, while size categories 4 and 5 were classified as big. None of the weeds were predominantly in the size 3 category. Therefore, classification of the weeds in category 3 was done by relating it to the immediate preceding and succeeding sizes of 2 and 4. For example, *Aristolochia bracteolata* occurs most significantly in size category 2, and in lower numbers in size category 3, therefore, the final size was determined to be small.

Size category 1: <0.425 mm (falling through Sieve C)
Size category 2: >0.425 mm but < 2 mm (staying in Sieve C)
Size category 3: >2 mm but < 4 mm (staying in Sieve B)
Size category 4: >4 mm but < 7 mm (staying in Sieve A)
Size category 5: > 7 mm (staying in ethnographic Sieve)

Headedness

The tendency of weed seeds to remain in heads, spikes, or clusters in spite of threshing is an important variable to consider in classifying the weeds. The tendency to remain in heads is a result of the degree of immaturity of the weed plants at the time of harvest. A high tendency for headedness results in distinct processing patterns. It is quite common that when weed seeds are headed they are less winnowable by wind, and are most effectively removed from the crop product through winnowing by shaking. Headed weeds are not always processed out during sieving because the ethnographic sieve is large enough for the headed seeds to pass through (the main purpose of the ethnographic sieve being to separate out larger non-crop components such as straw, inflorescence heads, stalks). In sum, headedness makes a weed seed larger and more visible for processing out during the winnowing by shaking, and subsequent stages.

The measurement of 'headedness' much like the next variable 'aerodynamics/weight' is a more subjective measure than that of weed sizes. 'Headedness' was classified on a scale of 1 to 5, where 1 was very likely to be free and 5 was very likely to be headed. The weeds were rated according to where they fell on the continuum, and this was decided primarily through comparative observations among them, and, in addition, observations on how they appeared in the processing stages; for example, if they continued to appear headed in the later stages, despite the threshing and sieving that could have dislodged their headed nature.

Weight / Aerodynamics

This characteristic is of importance in winnowing processes rather than sieving, though if seeds are particularly small and heavy they could pass through the sieve with more ease than if they were light and small. Additional appendages are of considerable importance in the determination of aerodynamics. However, there were only a few weeds in this assemblage that had aerodynamic appendages such as hairs, wings, or equivalent dispersal mechanisms.

This variable was measured by utilizing data on weight, shape, and aerodynamic appendages of the weeds. Therefore, it is a classification that is the result of the three measurements. Weed seed weight is important for evident reasons: the heavier the seed the less its aerodynamics and vice versa. Weed seed shape is considered significant because it affects the processor's discrimination due to any similarity to the crop seed. The higher the similarity in shape and size, then the greater the chances of such weed seeds not being processed until the later stages where hand sorting is predominant. The weed seeds are scored on a scale from 1 to 5, with 1 being very light and highly aerodynamic and 5 being very heavy and with low aerodynamics.

Conclusion

Using the three main characteristics presented in the above discussion, seed size, seed headedness and seed aerodynamics, the weeds from the ethnographic studies in India were classified and grouped into 8 categories to facilitate modeling of crop processing pathways:

1. Small Free Heavy (SFH)
2. Big Free Heavy (BFH)

3. Small Free Light (SFL)
4. Big Free Light (BFL)

5. Small Headed Heavy (SHH)
6. Big Headed Heavy (BHH)

7. Small Headed Light (SHL)
8. Big Headed Light (BHL)

Crop Types

The underlying premise of the ethnographic crop processing studies conducted in India was that each step of crop husbandry and grain processing has a measurable effect on the composition of crop products and byproducts. This same ethnographic research has confirmed that the wheat/barley/pulse models of crop processing developed by Hillman (1981, 1984) and Jones (1983, 1984, 1987) are not applicable for millet crops such as *Sorghum bicolor, Pennisetum typhoides,* and *Panicum miliare*. One of the major differences between these two groups of crops lies in the harvesting methods.

As was discussed in the previous chapter, millets crops such as *Sorghum bicolor, Pennisetum typhoides* and *Eleucine coracana* are harvested in several different ways. One of the distinctive methods, termed the Type I method, is by only cutting off the inflorescence heads and leaving the stalks behind. This initial variation in processing has a measurable effect on the composition of crop products and byproducts, primarily in the inclusion of weeds. Crops that can be harvested in this manner have been classified as Type A crops (see Figure 3-2).

Other millet crops such as *Setaria* spp. and *Panicum* spp. are termed as Type B crops (small grained plant crops such as *Chenopodium* spp. can also be included into this type). Their harvesting method includes the inflorescences heads, the stalks, and most often they are cut at the base of the plant (see Figure 3-3). This method, Type II, incorporates the weeds growing in between and around the crop plants into the processing stages. These weeds then appear in the crop products and have to be methodologically removed through various processing methods. Thus the composition of the different stage crop products and byproducts is markedly different from Type A Crop processing.

It is important to keep in mind the distinction between Type A and B crops, and Type I and II methods of harvest. Type A crops can on occasion be harvested as a Type B crop, but Type B crops are very rarely or almost never harvested as Type A crops (Table 4-1). Therefore, there is a higher probability that Type B crops products are more likely to contain weed seeds than Type A crops. Furthermore, when weed seeds occur in Type A crop products, then, depending on their frequencies, they are anomalous intrusives, or the Type A crop was harvested like Type B crop.

This distinction made between crop types based on harvesting method is important because this first crucial step in the processing has a critical effect on the crop product and byproduct composition of successive stages. It is therefore important to understand the effect its harvesting method has on the process for each crop, and for different ecological regions. When there is variation in crop product composition within and between different ecological areas (especially in terms of absence or low weed inclusions), it is important to determine the reasons for these variations. The harvesting method should be the first indicator, albeit initial, before pursuing explanations such as effects of processing methods or ecological factors.

Type A Crops: Summer Cultivation of *Sorghum bicolor* and *Pennisetum typhoides*

The objective of the ethnographic crop processing study is to isolate the variables that identify millet processing for human food, both as human food and animal fodder, or in rare cases as only for fodder. My ethnographic observations in Gujarat, India, have noted that millets are rarely grown solely as fodder, and it is only in very rare instances that millet grain is fed to animals as fodder. An analogous practice could have been present during the Harappan period. Typically the grain is used for human consumption, while the byproducts (the crop residues) are used as animal fodder. In some instances, typically where fuel sources are infrequent, millets are grown only as human food and the residues are used as domestic fuel and these are often the only source of fuel. This is generally the situation when the number of animals owned by the farmer is relatively small, and they can be supported by free-range grazing.

In Gujarat, the primary millets grown by both the farmers and pastoralists are *Sorghum bicolor* and *Pennisetum typhoides*. The crop byproducts from these crop harvests and processing are fed to cattle. Occasionally *Panicum miliare* is also grown as green fodder for cattle. Green fodder is preferred for dairy animals, due to its low fiber content and high protein.

As presented in chapter 3, the ethnographic crop processing studies of summer monsoonal cultivation of millet crops took place in the settled village of Babarkot in Bhavnagar District, Gujarat. The two millet crops chosen for the study were those of *Sorghum bicolor* (Jowar) and *Pennisetum typhoides* (Bajri), both of which are Type A crops (Reddy 1991a). This research has attempted to examine whether Hillman (1984) and Jones' (1987) 14 processing stages distinguished for free-threshing cereals and pulses are applicable to millet processing. During this study, emphasis was particularly placed on documenting the cultivation of millets and their ultimate use, and the research was designed to quantitatively discriminate the variables that correlate the cultivation of millets with their ultimate use.

Table 4-2. *Sorghum bicolor.* Crop Processing Stages Composition (Combined Plots).

Standardized Counts for *Sorghum bicolor* (per 500 gm for each stage)

	Th Prd	**Win(w) Prd**	**Win(w) Byp1**	**Win(w) Byp2**	**Siev Prd**	**Siev Byp**	**Win(s) Prd**	**Win(s) Byp**
Straw Vol (ml)	27623	1163	68170	288892	1213	4744	–	4333
Sorghum Rachillas w/spikelets	2134	21631	1214	3140	–	107632	–	–
Sorghum infl heads	328	43	813	556	–	157	–	–
Sorghum w/ husk	26622	41654	4946	–	23467	–	25743	106978
Sorghum grain	13882	38401	375	–	28928	2369	17924	7937
Sorghum rachis	–	–	–	–	121	–	–	–
Total Weeds	78551	98955	8495	30448	118139	4916	815	332848
SFH Weeds	4812	3555	984	5916	5531	2213	371	23728
SFL Weeds	41137	40729	1835	5080	54320	565	96	212703
SHL Weeds	1468	366	3814	9663	30	1939	–	81
SHH Weeds	31000	54216	1333	7870	58218	39	348	96295
BHL Weeds	16	–	181	979	–	13	–	–
BHH Weeds	8	–	–	–	–	–	–	–
BFL Weeds	110	348	348	940	40	147	–	41

Th Prd = Threshing Product; **Win(w) Prd** = Winnowing by wind Product; **Win(w) Byp1** = Winnowing by wind Byproduct 1; **Win(w) Byp2** = Winnowing by wind Byproduct 2; **Siev Prd** = Sieving Product; **Siev Byp** = Sieving Byproduct; **Win(s) Prd** = Winnowing by shaking Product; **Win(s) Byp** = Winnowing by shaking Byproduct

SFH = small free heavy; SFL = small free light; SHL = small headed light; SHH = small headed heavy; BHL = big headed light; BHH = big headed heavy; BFL = big free light

The study plots selected in the fields of *Pennisetum typhoides* were three 5 by 5 meter squares, and the plots in the *Sorghum bicolor* fields were two 5 by 5 m squares (see Figure 3-5). The plots were characterized as follows: P1 (close to a water channel, cattle shed, pathway, and field of cotton and *Sorghum bicolor*); P2 (most isolated, furthermost from cattle, pathway, another crop field); P3 (furthermost from water channel and cattle), S1 (closer to water channel, cattle, pathway and another crop field); and S2 (further from water channel, cattle, pathway, and another crop field) (see Figure 3-5). The importance of these criteria is twofold: the variation of weeds and other components could be correlated to these different variables; secondly, and more importantly, having plots which satisfy all these different conditions provides a more representative sampling of the field for the study.

Sorghum bicolor

Sorghum bicolor is an annual with an erect solid stem and a compact inflorescence head at the top of the stem (Figure 4-1). The compact panicle has an axis or rachis of variable length (Cobley 1956) (Figure 4-2). *Sorghum bicolor* has paired spikelets borne on rachii at the tips of branched stiff rachillas. The maturity of the inflorescence head starts at the apex and continues downward. The caryopsis is rounded and varies in shape and size. It is comparable to naked wheats, since it

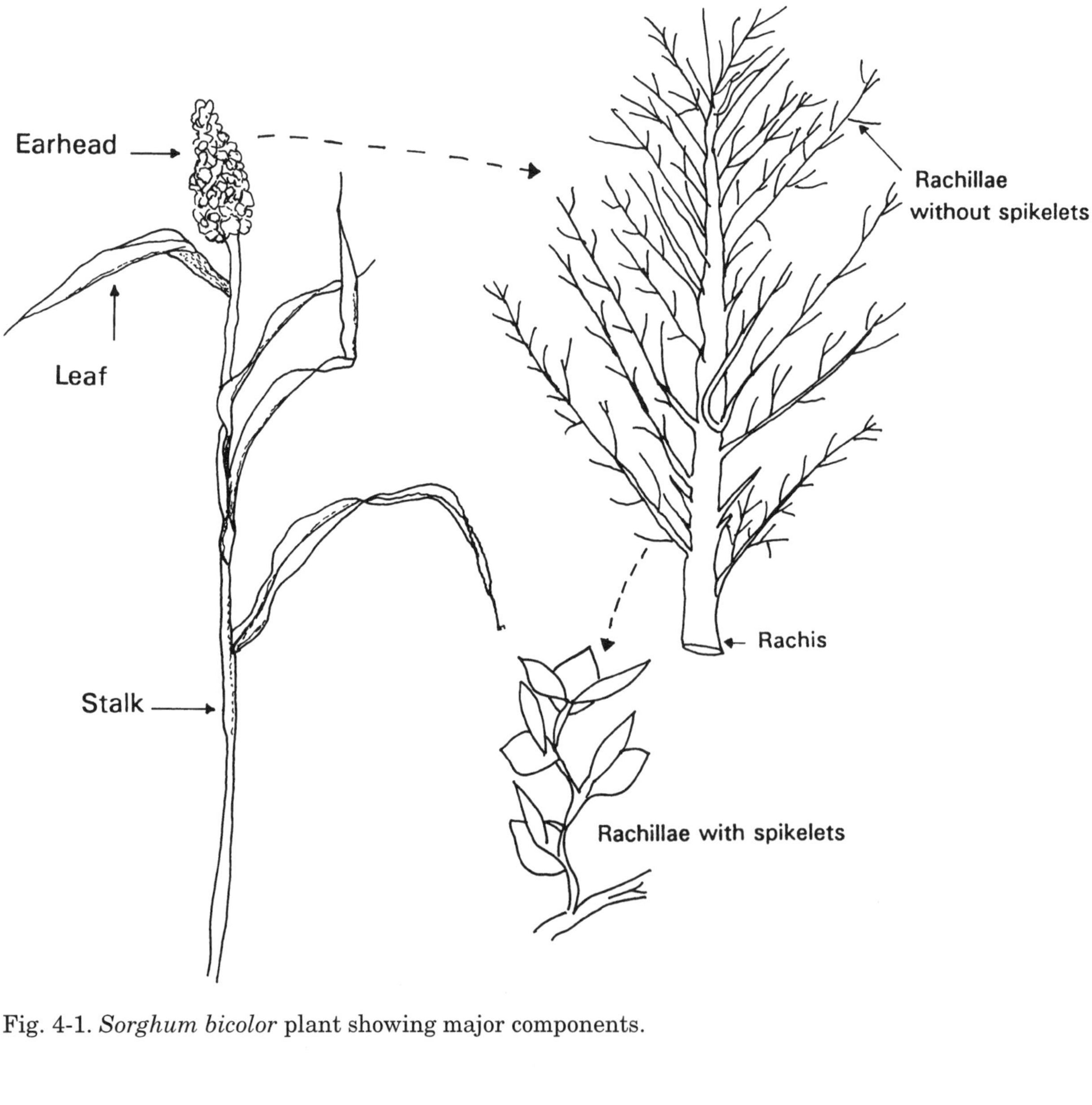

Fig. 4-1. *Sorghum bicolor* plant showing major components.

comes free of the palea and lemma without much threshing, although occasionally one does not get freed. The major components identified during the composition analysis of the *Sorghum bicolor* samples are: inflorescence heads, straw, chaff, *Sorghum bicolor* seeds with husk, *Sorghum bicolor* seeds without husk, *Sorghum bicolor* rachillas with spikelets, and a range of weed species.

The data collection and analysis was a two-step process. During the first step, a subsample was examined and by sorting through the materials the different categories were identified based on plant morphology. It is assumed that the categories most distinguishable from the caryopsis are processed out earlier in the processing stages, while those that are most similar to the caryopsis are processed in the later stages.

The second step involved the recording and documentation the frequencies, weights, and volumes of these categories (wherever appropriate) (Table 4-2). As a result of this composition analysis, several general observations can be made of the *Sorghum bicolor* samples.

General Patterns

Significant differences were observed between the two *Sorghum bicolor* plots, S1 and S2. The main

Fig. 4-2. *Sorghum bicolor* inflorescence head.

differences lie in the amounts of straw, inflorescence heads (panicle heads), and the types of weeds.

Straw is present in both plots but not in all the processing stages of the two plots. In plot 1 straw was processed out efficiently by winnowing by wind and sieving, but in plot S2 the straw was not processed out until the next stage down the line, i.e., winnowing by shaking. This is indicated in the data with straw absent in the winnowing by shaking product of plot S2 but present in the previous stages (sieving product, winnowing by shaking product, and winnowing by shaking byproduct) but only in sieve C. In plot S1 straw is absent in sieving product, winnowing by shaking product, and winnowing by shaking byproduct. The efficiency of the processor is not the reason for these differences since the same person processed both study plots. The most likely explanation lies in the fact that plot S2 was drier at the time of harvest, due to its greater distance from a water channel. Therefore, since it was drier there was more breakage of the plant material (straw in this case) during threshing by sticks. This resulted in the production of very small sized straw, and this sieve C straw was not being processed out in the earlier stages due to its small size.

The other difference between the two *Sorghum bicolor* plots lies in the distribution of inflorescence heads between stages. The inflorescence heads occur in the threshing product, winnowing by wind product, winnowing by wind byproduct 1, and sieving byproduct in both the plots. In addition, the inflorescence heads also occur in the winnowing by wind byproduct 2 of plot S2. This assemblage, winnowing by wind byproduct 2, is the lighter of the two byproducts and is often processed through several stages to retrieve the crop grain in the byproduct. The presence of the inflorescence heads in this assemblage of plot S2 is a direct result of the inflorescence heads being immature, and also the smaller sized ones being lighter and getting included into this byproduct. The presence of these smaller inflorescence heads in plot S2 may also be because this plot was further away from the water channel, and this had a direct effect on the growth and maturity of the inflorescence heads.

The last difference between the two plots entails variation in the distribution of weeds between processing stages. Specific weeds found in one plot are absent in the other, and vice versa. The weeds were categorized to determine if there were any patterns of occurrence according to the categories, but there was no correlation. The most important observation is that weeds decrease significantly through the processing pathway and there is a definite patterning in their occurrence that can be used as indicators of different processing stages.

Finally, *Sorghum bicolor* seeds with husk are present in relatively high frequencies through the processing stages, and indicate that the processing is not a simple or arbitrary elimination activity, but a process where conscious attempt is made to increase yields, and to maximize the retrieval of the *Sorghum bicolor* grains both with and without husk during processing.

The processing pathway of *Sorghum bicolor* is not a simple linear route, but it is an involved process in which decision-making plays an important role (Figure 4-3). This decision-making is based on the experience of the processor, and also general factors such as amount and nature of the plant material being processed, economic needs, and cultural systems. For example, deciding whether to select a winnowing by wind byproduct for further processing, or to use it as fodder. Also, a decision must be made on when a byproduct can be used as fodder immediately, or when it is adequately large in quantity to store as for later use.

Detailed Analysis

After the samples were sorted and categorized, the data were analyzed to identify patterns that relate to the different processing activities and to isolate variables that are indicative of particular processing activities.

In the analysis process, the first step was to standardize the data to facilitate rigorous comparison between all the plots (including data from different crops such as *Panicum miliare* and *Pennisetum typhoides*). The standardization was based on weight rather than volume since volume measurements of some of the ethnobotanical materials such as straw and chaff are unreliable and inaccurate.

The data from both the *Sorghum bicolor* plots S1 and S2 were combined for analysis. They were combined to provide a more representative data set, since all the ecological variables identified in the field are represented in the two plots. Additionally, since there were no major unexpected differences between the plots, apart from the ones already explained in the previous discussion, combining the two data sets was appropriate. The results of the analysis and pattern searching are presented in the following discussion.

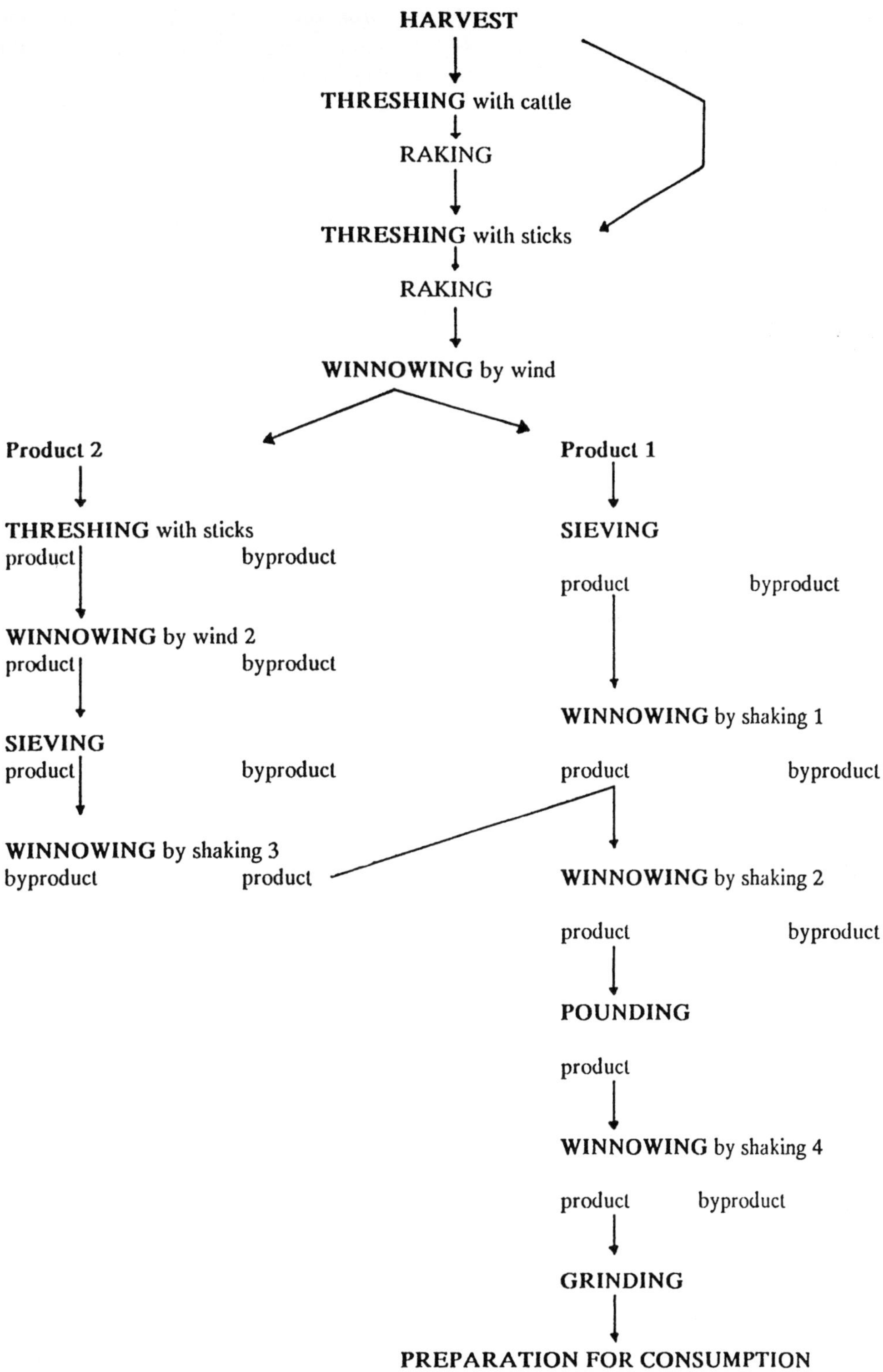

Fig. 4-3. *Sorghum bicolor* processing pathway.

Straw

The straw component is present in significant quantities in threshing product, winnowing by wind byproduct 1 and byproduct 2 (Table 4-2). The assemblages from these stages are also characterized by dominant frequencies of inflorescence heads. These two components are not considered strong isolating variables, since their archaeological preservation and discovery are quite unlikely. Straw also occurs in moderate amounts in winnowing by wind product, sieving product, sieving byproduct, and winnowing by shaking byproduct.

Winnowing by shaking process removes all remaining straw from the crop product, which was not processed out in any of the earlier processing stages (Table 4-2). Observations on the relationship between sieve size and the straw component patterning reveal that as the processing advances, straw in sieve C increases. In other words, in the later processing stages there is no sieve A straw and most straw occurs in sieve C. This pattern is expected since the processor is more likely to remove larger undesirable components in the earlier stages and the smaller ones later.

Inflorescence Heads

Distribution of the *Sorghum bicolor* inflorescence heads reveals that they dominate the winnowing by wind byproducts 1 and 2, and the threshing product (Table 4-2). They mostly occur in sieve A, but they only occur in sieve B of the winnowing by wind byproduct 2 from plot S2. The inflorescence heads occur in moderate amounts in sieving byproduct, and in much lower numbers in the winnowing by wind product.

The implication of this pattern is that the *Sorghum bicolor* inflorescence heads are processed out in significant numbers by the winnowing by wind process, and the remaining small quantity is processed out through sieving. Thus, *Sorghum bicolor* inflorescence heads can be used as a variable to isolate the lower processing stages such as threshing and winnowing by wind.

Sorghum bicolor *rachillas with spikelets*

This distinct component occurs dominantly in the sieving byproduct (Table 4-2). It occurs in moderate to low frequencies in the winnowing by wind product, threshing product, and winnowing by wind byproducts 1 and 2. It is absent in the sieving product, and winnowing by shaking product and byproduct. This patterning shows that the *Sorghum bicolor* rachillas with spikelets are quite discriminating in their occurrence and could be used as a relatively strong distinguishing variable, particularly for the sieving byproduct stage.

Sorghum bicolor *seeds with husk*

This component occurs in very dominant numbers in winnowing by shaking byproduct (Table 4-2). It occurs in dominant frequencies in threshing product, winnowing by wind product, sieving product, and winnowing by shaking product. Crop seeds with husks dominate the crop products of different processing stages showing that they are being chosen purposefully despite their husk (these grains will have to be dehusked specifically later). The dominant frequencies in the winnowing by shaking byproduct, however, are due to removal of seeds that are defective and infected. They are processed out only during the hand sorting activity in the winnowing by shaking process. This patterning also reveals an attempt to limit loss of *Sorghum bicolor* grain since the processor is making a strong effort not to remove the seeds with husk, but to invest labor in dehusking them at a later stage (but before the seeds are to be pounded or ground). These *Sorghum bicolor* seeds with husk are often then separated from the dehusked ones, and then husked through light pounding followed by winnowing by shaking to separate the chaff from the seeds. These seeds are then combined with the other *Sorghum bicolor* dehusked seeds.

Sorghum bicolor *seeds without husk*

The seeds without husk occur in varying frequencies in all the stages crop products and byproducts except winnowing by shaking byproduct 2 (Table 4-2). They occur in lesser frequencies in winnowing by wind byproduct 1 and sieving byproduct. As expected they occur in highly dominant frequencies in threshing product, winnowing by wind and winnowing by shaking products, and sieving product. Winnowing by shaking byproduct also has them in moderate amounts, as a result of intentional removal by shaking and hand sorting rather than accidental processing loss. Their presence in such moderate amounts is important, since the winnowing by shaking occurs in domestic contexts where the index of archaeological discovery and preservation is relatively high.

Weeds

Weeds in the *Sorghum bicolor* plots are present in relatively high frequencies in all the stages products and byproducts, except in sieving byproduct and winnowing by shaking product (Table 4-2). The weed categories present are small headed heavy (SHH), small headed light (SHL), small free light (SFL), small free heavy (SFH), bigheaded heavy (BHH), big headed light (BHL), and big free light (BFL). Only the weed category of big free heavy (BFH) is not represented. Of all the categories, the small free heavy (SFH) weeds are the most ubiquitous, followed by the small headed heavy (SHH) and small headed light (SHL). The big sized weed categories such as the big headed heavy (BHH), big headed light (BHL) and big free light (BFL) occur in low frequencies, while the small sized weeds have significantly higher in frequencies. The ratio of headed to free weed seeds is equal showing that both types of weeds occur in the processing stages, and the key question is where they drop off from the crop products within the pathway.

The patterning of the weed categories down the processing pathway is intricate in that different categories occur in differing numbers at different stage products and byproducts. Weed categories are processed out at different stages due to their morphological characteristics. For example, although most weeds are processed out by winnowing by shaking, some of the SHH, SFH and SFL category weeds remain, albeit in extremely low frequencies, in the winnowing by shaking 1 product. The SHL weed category, however, is primarily processed out by sieving, and the remaining SHL weeds are removed by the winnowing by shaking process. The BHL category of weeds is processed out first in limited numbers through winnowing by wind, and then primarily by sieving. The BFL weeds are processed out through sieving. Thus, distinct weed categories occur in varying levels of significance at different processing stages. Therefore, they can be used as indicators of these processing stages' crop products and byproducts. It is important to emphasize that they alone cannot be used as discrete isolating factors. Instead, a combination of specific weed category ratios and other components, such as crop grains with or without husk, or rachillas with spikelets, must be used to successfully identify assemblages relating to the crop products or byproducts of different processing stages.

Summary of Sorghum bicolor

In summary, several strong observations can be made regarding the distribution of the various components in the crop products and byproducts. These observations have important implications for modeling the discrimination of crop processing stages for archaeological interpretations. *Sorghum bicolor* inflorescence heads are definite indicators of lower processing stages such as threshing and winnowing by wind, but they cannot be used as the exclusive isolating variables for archaeological contexts since they are unlikely to preserve in the archaeological record. For archaeological contexts, the preservation of panicle heads and their discovery are mainly dependent on whether they are carbonized or not. There are two critical factors that determine whether this component is archaeologically important as a strong isolating variable: the location of the processing activities and the likelihood of the components being exposed to fire. Threshing and winnowing by wind most often occur in fields/non-domestic contexts which are unlikely to be discovered archaeologically, and their preservation in such contexts through exposure to fire is unlikely. Thus, the *Sorghum bicolor* inflorescence heads, though good indicators of threshing and winnowing by wind, have a lower index of significance as an isolating variable (Table 4-2).

Sorghum bicolor rachillas with spikelets can be used as distinguishing variables, since they occur in dominant frequencies (as compared to other components) only in the sieving byproduct (Table 4-2). Clearly, rachillas with spikelets were being processed out by sieving, resulting in their high frequencies in the sieving byproduct. They occur in low and moderate frequencies in the threshing and winnowing by wind assemblages, and are absent in the sieving product and winnowing by shaking assemblages. Therefore, an assemblage dominated by *Sorghum bicolor* rachillas with spikelets is most akin to a sieving byproduct assemblage.

Sorghum bicolor seeds with or without husks are important indicators of crop products from all the stages (Table 4-2). When seeds dominate an assemblage, then it can be certainly interpreted as a crop product. However, *Sorghum bicolor* seeds with or without husk alone cannot be used to determine the particular processing stage of the crop product. In conjunction with other components, the stage can be distinguished. For example, if the relative proportions of *Sorghum bicolor* seeds (husked or/and dehusked) are much higher than other components, ranging from the majority to total dominance, then the assemblage could be interpreted as representing a winnowing by shaking 1 or 2 product. Once again, it is important to emphasize that typically one component cannot be used as a singular isolating indicator or variable.

Weed categories are also important distinguishing variables. In conjunction with other components they provide signatures of assemblages relating to different stages' products and byproducts. It is important to consider how the different weed categories interplay with the different processing activities. For example, the headedness of a weed has a direct effect on whether it is sievable or not, whether it is winnowable by wind or by shaking, and so on. Similarly, the size and the weight have direct effect on where and how the weeds are processed.

The final observation relates to the distribution of components between sieve sizes. In general, the higher the processing stage (closer to the consumption product) the fewer sieve A components in the crop products and byproducts. In the last few stages, such as winnowing by shaking 1, winnowing by shaking 2, pounding and winnowing by shaking 3, the crop products and byproducts have only sieve B and sieve C components, with the majority being in the sieve C. *Sorghum bicolor* seeds are never in sieve A, instead they are always in sieves B and C. This pattern is expected since the bigger the size of non-crop grain components (sieve A sized components), the higher the chances of them being processed out. They are simply more visible and therefore easier to hand sort out and also easier to winnow out through wind (especially straw). The use of the sieve at a relatively early processing stage further facilitates the separation of the bigger sized material.

Pennisetum typhoides

Pennisetum typhoides is an annual grass, with a solid erect stem, which bears the compact inflorescence at the tip (Figure 4-4). The plant tillers profusely and produces a leafy growth. The unbranched inflorescence has a straight rachis, which bears the spikelets on an involucre and rachillas (Figure 4-5). These spikelets bear the seeds that are caryopsis. The rachis is unbranched, solid and straight, tapering gradually from base to apex. *Pennisetum typhoides* caryopsis do not have a clasping palea and lemma similar to free threshing wheats. Therefore, they thresh free without much effort on the part of the processor, although there are the occasional ones that do not get freed. During the composition analysis the major components isolated in *Pennisetum typhoides* samples are: straw, husk, *Pennisetum typhoides* rachii with rachillas bearing spikelets, *Pennisetum typhoides* rachii without rachillas, *Pennisetum typhoides* rachillas without seeds, *Pennisetum typhoides* rachillas with seeds, *Pennisetum typhoides* seeds, *Pennisetum typhoides* seeds with husk, and a range of weed species (see Figures 4-4 and 4-5).

General Patterns

Differences were observed between the three *Pennisetum typhoides* plots with the main differences being the frequencies of the rachillas without spikelets, and the ranges and frequencies of weed categories. *Pennisetum typhoides* plot 3 has a significantly higher frequency of rachillas without spikelets, while the *Pennisetum typhoides* plots 1 and 2 have lower and comparable frequencies of this component. This pattern is most likely a result of plots 1 and 2 being relatively close to a water channel, while *Pennisetum typhoides* plot 3 was much further from a water channel. Less water availability in plot 3 plants resulted in plants being dry well before harvest and threshing. Thus, it was easier for the spikelets and grain of drier *Pennisetum typhoides* panicle heads to dislodge from the rachillas, resulting in the higher frequencies of rachillas without spikelets in plot 3.

Overall, there were fewer weeds in *Pennisetum typhoides* plot 3 as compared to plots 1 and 2. Plot 1 has the highest frequency of weeds but it is also very disturbed in terms of it being the closest to a pathway, and also to a field of a different crop (field weeds are most often associated with disturbed areas). These factors could have an effect on the numbers of weeds. However, plot proximity to a water channel could also be a causal factor in this patterning. Both plots 1 and 2 have water channels in relatively close proximity as compared to plot 3 (see Figure 3-5). Therefore, it is possible that plot 3 has a lower frequency of weeds because it was getting less water and was drier, while plots 1 and 2 have a higher frequency of weeds because they were close to disturbance areas (pathway and other crops). The differences could also be due to sample size and variability, and not related to any ecological factors.

The processing pathway of *Pennisetum typhoides* is a complex nonlinear procedure where decision making on the part of the processor plays an important role (Figure 4-6). This decision-making is based on experience and other factors such as amount of material being processed, economic needs, and cultural values. For example, deciding that a winnowing byproduct has significant amount of crop grain that it cannot be discarded; instead a separate processing sub-pathway should be initiated to extract the grain. This decision-making is not quantifiable, but is a very important aspect of the processing.

A second observation regarding *Pennisetum typhoides* processing relates to how weeds fall out in the processing pathway. Weeds occur in higher frequencies in the assemblages of the later processing stage products and byproducts (as com-

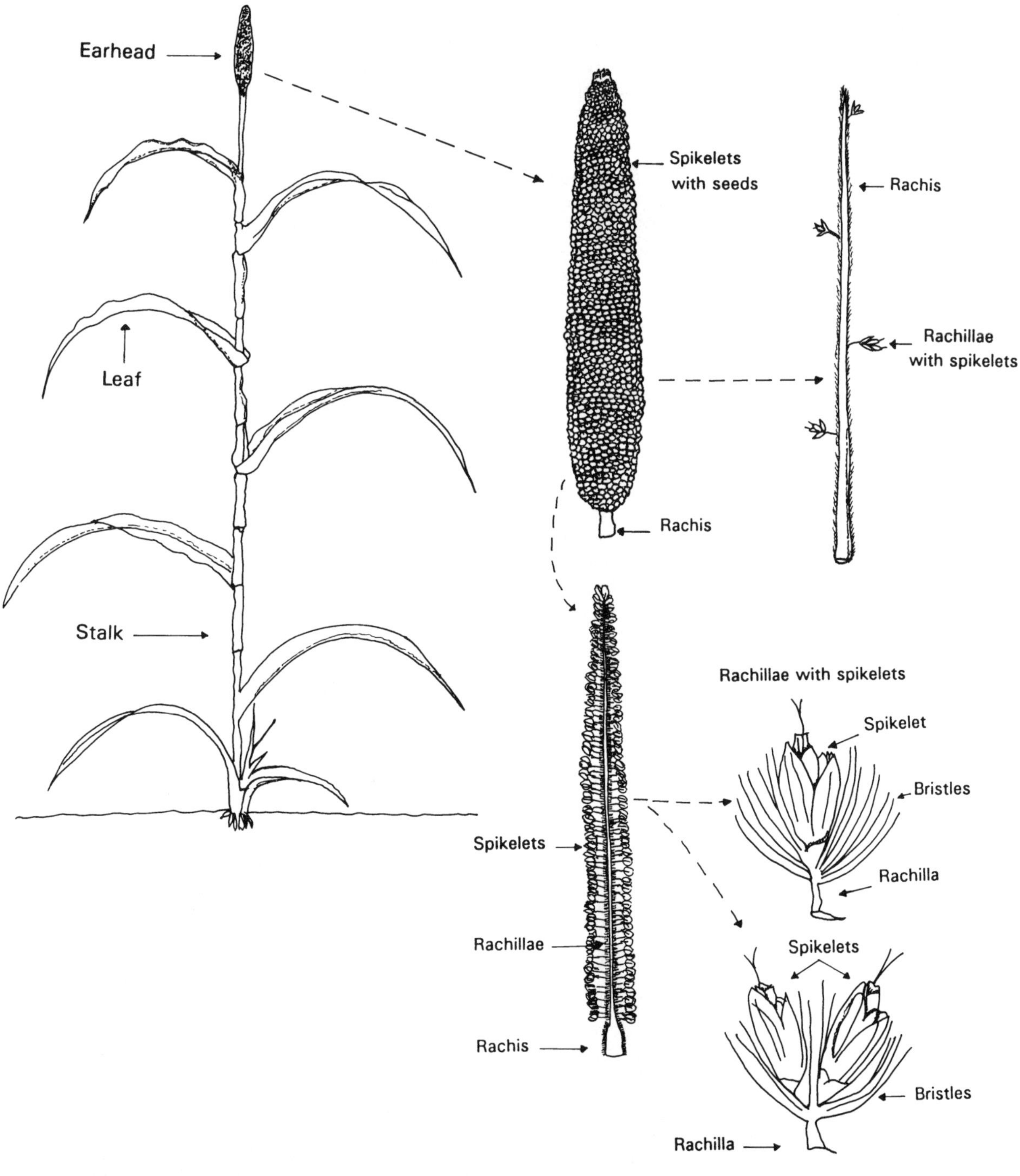

Fig. 4-4. *Pennisetum typhoides* plant showing major components.

pared to the earlier ones). Essentially, they are actively processed out in the later stages in significant frequencies. This differs quite distinctly from *Sorghum bicolor* processing where weed frequencies decrease significantly through the processing pathway. One reason for these differences could relate to the different sizes of the two crop grains, particularly in relation to the weed seed sizes, even though both *Sorghum bicolor* and *Pennisetum typhoides* are Type A crops. The *Sorghum bicolor* seed is much bigger than the *Pennisetum typhoides* seed. Therefore, the size of the *Pennisetum typhoides* crop grain is effecting the removal of other seeds (weeds) in the initial stages since the focus of these initial processing activities is to separate the larger non-crop grain components such as straw, rachii, and rachillas. In the later stages when the focus is more on the smaller sized components, the weeds are processed. Therefore, the smaller the prime grain the greater the number of weeds in the later processing stages. Furthermore, in general there is purposeful concentration on processing out larger sized non-crop grain components in the earlier stages; therefore, if an assemblage (archaeological or ethnographic) has mostly crop grain, smaller non-grain crop components and weeds, it

Fig. 4-5. *Pennisetum typhoides* inflorescence head.

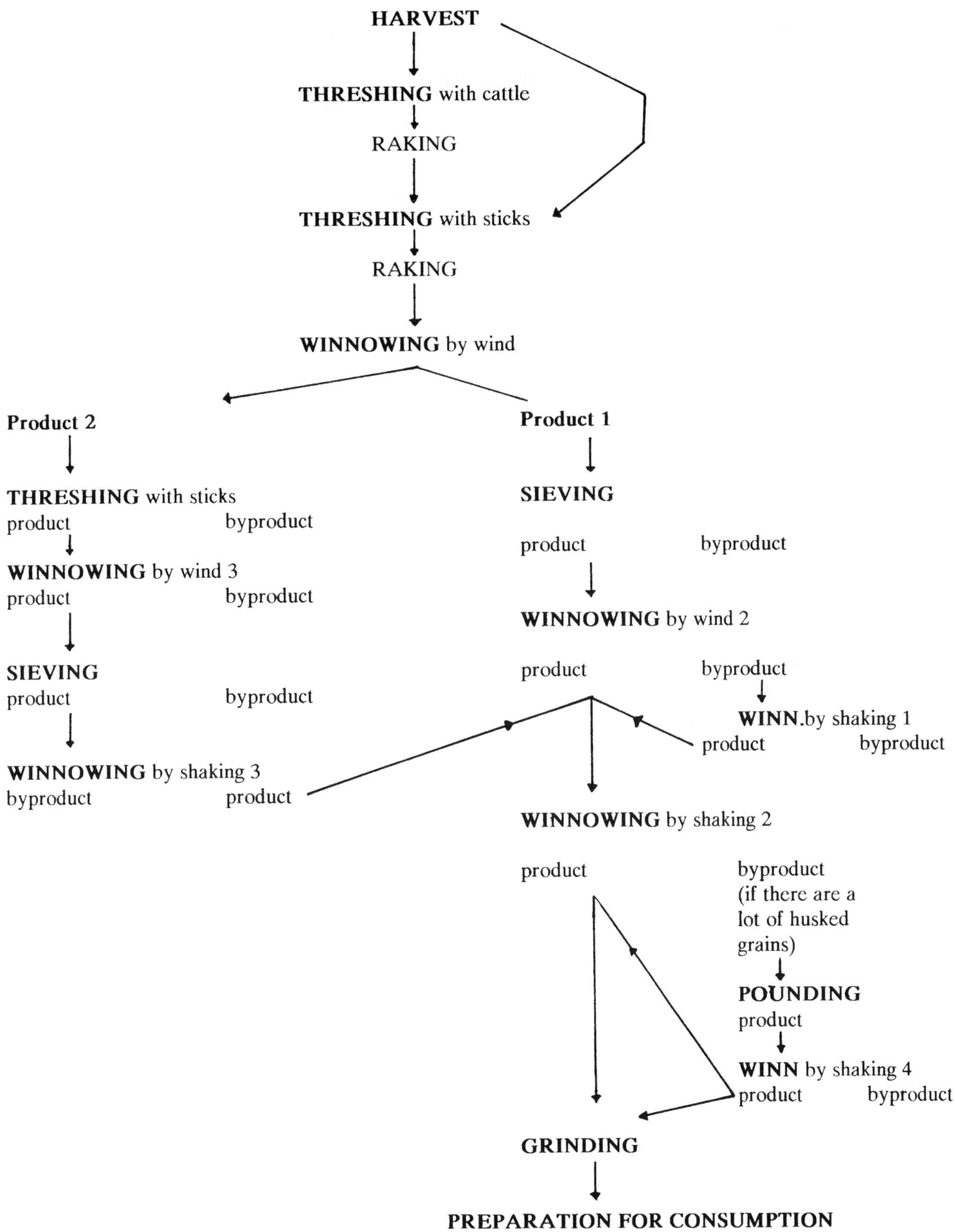

Fig. 4-6. *Pennisetum typhoides* processing pathway.

is most likely closely related to a later processing stage product than an earlier stage such as winnowing by wind or sieving.

Another observation related to the patterning of the weeds is that there is a significant loss/elimination of tail grain in later stages such as winnowing by wind 2 byproduct, and winnowing by shaking byproduct. There is relatively little loss of *Pennisetum typhoides* grain (tail and prime) in the initial stage byproducts. This changes dramatically in the later stages when there is a conscious attempt on the part of the processor to eliminate the tail grains.

Lastly, similar to *Sorghum bicolor*, *Pennisetum typhoides* grains with husk occur in relatively high percentages throughout the processing stages. I propose that this patterning reveals that processing is not a simple elimination activity where there is singular selection for *Pennisetum typhoides* prime grains without husk, but rather is an involved process where there is a conscious effort to maximize the retrieval of *Pennisetum typhoides* seeds, both with and without husk.

Detailed Analysis

After sorting and categorization, the data analysis involved identification of patterns that related to the different processing activities and isolation of variables that are indicative of particular processing activities.

Standardization was the first step of the data analysis. As mentioned previously, this was done using standardized weights to make the data from *Pennisetum typhoides* samples comparable to the *Sorghum bicolor* and *Panicum miliare* data. The standardized data were then used for pattern searching and analysis (Table 4-3).

Data from the three plots of *Pennisetum typhoides* were averaged to create a more representative sample encompassing the range of ecological variables identified in the field. The analysis and pattern searching of crop components were aimed at isolating variables that distinguish products and byproducts of various processing stages.

Straw

The straw component is present in significant frequencies primarily in the initial stages of processing, and when present in the later stages, it only occurs in sieve C. Straw is present in relatively significant amounts in the threshing products, winnowing by wind product and byproduct, and sieving byproduct (Table 4-3). In winnowing by wind 2 byproduct, it is present only in sieve C. These stages are also characterized by high frequencies of rachii, both with and without rachillas. These two components, straw and rachii, are distinguishing variables for these stages, but they are not considered to be strong isolating variables for archaeological modeling since their preservation, survival, and discovery are unlikely. This is due to their characteristics and nature, and the processing stage locations. Straw is processed out completely in the sieving that follows winnowing by wind. It was observed that as the processing stage advances, the straw is primarily limited to sieve C. This pattern supports the assumption that the processor and the processing activities first select out the larger material in the earlier stages and the smaller materials are selected out in the later stages.

Rachii

Analysis of the distribution of *Pennisetum typhoides* rachii without rachillas reveals that they are most prevalent in sieving byproduct, followed by the winnowing by wind byproduct and product (Table 4-3). Most are processed by sieving, which is later than winnowing where there is some processing out of the rachii without rachillas. Similarly, rachii with rachillas are most commonly found in the sieving byproduct, followed by winnowing by wind byproduct and product. Thus, both rachii without rachillas and rachii with rachillas are partially processed out in the winnowing by wind and then intensively processed in sieving. This pattern in expected since the rachii (with or without rachillas) are heavy enough not to be processed out fully by winnowing by wind, but are large enough to be sieved out in the ethnographic sieving. These are large components relative to the prime grain, and are processed out in the earlier stages, and their presence archaeologically can be used as a distinctive indication of an earlier stage of processing.

Rachillas without spikelets

The next component, rachillas without spikelets, are dominant in the first winnowing by wind byproduct (Table 4-3). This component is present to some degree, even if in relatively negligible percentages, in all the stages studied. The relative frequency of the component decreases with each processing stage, and the majority are processed out through winnowing by wind. Their presence is strong in the threshing products and moderate in sieving byproduct, but in the rest of the stages' product and byproducts they occur in very low percentages. Therefore, when rachillas without spikelets dominate an assemblage, the assemblage can be considered as being closely related to a winnowing by wind byproduct.

Rachillas with spikelets

Rachillas with spikelets also occur in all stages' products and byproducts (Table 4-3). However, they are most frequent in sieving byproduct, and occur in very low percentages elsewhere. In sieving byproduct, they occur moderately and in similar frequency to rachillas without spikelets. Although

Table 4-3. *Pennisetum typhoides.* Crop processing ptages composition (combined plots).

Standardized Counts per 500 g for each stage

	Th Prd 1	**Th Prd 2**	**Win(w) Prd**	**Win(w) Byp**	**Siev Prd**	**Siev Byp**	**Win(w) 3 Prd**	**Win(w) 3 Byp**	**Win(s) Byp**
Straw (ml)	357	2345	16	16412	–	1112	–	–	–
Rachillas w/o Seeds	28348	173340	806	661407	14	31715	–	666	129
Rachillas w/seeds	2747	1624	2349	465	15	34711	–	366	43
Rachii w/o rachillas	–	98	2	11	–	85	–	–	–
Rachii w/rachillas	7	146	4	162	–	182	–	–	–
Penn spikelets	2019	826	1787	624	1551	5541	971	6036	3974
Penn. Grain	62269	54227	47873	829	50861	27521	54040	74594	81049
Total Weeds	2317	2923	2606	4617	1791	1767	5379	51277	11094
SFH Weeds	1145	722	1179	1020	783	646	2475	19947	3119
SHH Weeds	376	97	358	281	286	23	1225	3003	1762
SFL Weeds	640	1044	924	469	715	54	1629	26720	5065
SHL Weeds	105	920	129	1908	1	804	16	1568	926
BFL Weeds	50	138	13	928	4	193	32	39	177
BHH Weeds	1	1	3	–	4	39	2	–	26
BHL Weeds	–	1	–	11	–	–	–	–	19
BFH Weeds	–	–	–	–	1	8	–	–	–

SFH = small free heavy; SHH = small headed heavy; SFL = small free light; SHL = small headed light; BFL = big free light; BHH = big headed heavy; BHL = big headed light; BFH = big free heavy

Penn = *Pennisetum typhoides*

Th Prd 1 = Threshing Product 1; **Th Prd 2** = Threshing Product 2; **Win(w) Prd** = Winnowing by wind Product; **Win(w) Byp** = Winnowing by wind Byproduct; **Siev Prd** = Sieving Product; **Siev Byp** = Sieving Byproduct; **Win(w) 3 Prd** = Winnowing by wind 3 Product; **Win(w) 3 Byp** = Winnowing by wind 3 Byprod; **Win(s) Byp** = Winnowing by shaking Byproduct they do not dominate the sieving byproduct, when an assemblage has moderate numbers of rachillas with spikelets, along with moderate numbers of rachillas without spikelets, then the assemblage can be correlated with a sieving byproduct. Rachillas with spikelets, however, are not as strongly associated with one stage as are the rachillas without spikelets.

Pennisetum typhoides *grain with husk*

This component occurs in all the products and byproducts in varying frequencies, but is dominant in winnowing by wind 2 byproduct and sieving byproduct (Table 4-3). Its presence throughout the pathway reveals that there is limited purposeful selection against this component, and that the *Pennisetum typhoides* grain with husk is desirable grain. Their relatively high frequencies in winnowing by wind 2 byproduct and the sieving byproduct do not contradict this assertion, since the grains with husk which occur in these two byproducts have distinct set of characteristics that make them undesirable grain. A significant percentage of the *Pennisetum typhoides* grain with husk in the winnowing by wind 2 byproduct are tail grains recovered from sieve C (prime grains are predominantly in sieve B). The grains in sieving byproduct are predominantly the fungal infected grains. Fungal infection causes them to become enlarged and they are sieved out easily. Tail grains are selected out purposefully since they affect the final quality of the crop grain. Concurrently, it is evident that the processor is making a conscious attempt not to remove the *Pennisetum typhoides* grains with husk, instead to invest in dehusking them at a later stage (but before the grains are pounded or ground into flour). Depending on the frequency of grains with husk, they can be separated from the grains without husk, pounded to dehusk them, and then recombined before grinding them all together into flour.

Pennisetum typhoides *grains without husk*

The grains without husk are the prime grain and the most desirable component. They are present in varying frequencies in all the stages' crop products and byproducts. They dominate the assemblages of winnowing by wind product, sieving product, winnowing by wind 2 product, and winnowing by shaking byproduct. They occur in relatively strong proportions in the threshing products and winnowing by wind 2 byproduct. The byproducts of sieving and winnowing by wind have moderate to low amounts. It is expected that the products of the different stages have *Pennisetum typhoides* grain without husk in dominant frequencies (the presence of grain in winnowing by shaking byproduct, however, needs further explanation). Sieve size distribution is of crucial importance in understanding the distribution of grain in products versus byproducts. Winnowing by shaking byproduct has *Pennisetum typhoides* grain without husk in dominant frequencies, but they are mostly of sieve C size (therefore they are interpreted as tail grain). In contrast, the various stage products have predominantly sieve B grain, with the sieve C grains decreasing as the stage product is higher in the processing pathway (in other words closer to the consumption product). Thus, the size of the *Pennisetum typhoides* grains plays an important role in determining when an assemblage is a byproduct versus a product. Additionally, the tail grains are processed out predominantly in the later winnowings, since the *Pennisetum typhoides* grain without husk sieve C occur in high frequencies mainly in the winnowing by wind 2 and winnowing by shaking byproducts. This pattern again demonstrates the concentration on the part of the processor during of the earlier processing stages on the elimination of larger non-crop grain components, while in the later stages the focus moves to the smaller undesirable components. This pattern has important implications for archaeological modeling, since the probability of the later stages' products and byproducts surviving and being discovered is much higher, as compared to the earlier stages, since the later processing stages occur almost predominantly in domestic locales.

Weeds

Weeds in *Pennisetum typhoides* plots are present in relatively high frequencies in all the stage products and byproducts studied (Table 4-3). All the weed categories are represented and consist of small headed heavy (SHH), small free heavy (SFH), small free light (SFL), small headed light (SHL), big headed heavy (BHH), big free heavy (BFH), big headed light (BHL), and big free light (BFL). The small free heavy weeds are the most ubiquitous, followed by small headed light (SHL), small headed heavy (SHH), and small free light (SFL). Not only do the small free heavy (SFH) weeds occur in the highest frequencies overall, and they are also dominant in most stage products and byproducts. They occur in higher frequencies in the various products, and do not get processed out till the later stages, particularly winnowing by wind 2 byproduct onward. In general, the four big sized weed categories occur in lower frequencies as compared to the smaller sized categories. Among the big sized weeds, the big headed heavy and big free heavy are the most frequent. The ratio of headed to free weeds is 1:1.3, meaning there is slightly higher incidence of the free characteristic weeds in the processing pathway than the headed weeds. The critical question, however, is where they are processed out from the crop products within the pathway.

The patterning of the weed categories down the processing pathway is intricate in that the different categories occur in varying frequencies at different stages. Small headed heavy (SHH), small free heavy (SFH), small free light (SFL) weeds, however, are processed out essentially by winnowing by wind 2, although they are also processed out minimally in the earlier stages. The small headed light (SHL), big headed heavy (BHH), big free heavy (BFH), and big free light (BFL) weeds are processed out by sieving. The big headed light (BHL) weeds are processed out mostly by the first winnowing by wind. Thus, different weed categories occur at differing levels of significance in the different processing stages. Therefore, they can be used as indicators of these processing stages' crop products and byproducts. Nonetheless, the weed categories alone cannot be used as isolating variables, but a combination of the ratios of specific weeds to at least one or more other components (such as crop grains with or without husk, rachii, rachillas) can be successfully used to identify assemblages relating to the crop products and byproducts of different processing stages.

Summary of Pennisetum typhoides

In summary, several key observations can be made about the patterning of various components in the products and byproducts from the different crop processing stages of *Pennisetum typhoides*. *Pennisetum typhoides* rachii with and without rachillas are clear indicators of lower processing stages such as threshing, winnowing by wind, and sieving (Table 4-3). They are predominantly processed out through sieving. However, they have limited utility for use as exclusive isolating variables for archaeological contexts, since they are not likely to survive into the archaeological record. For archaeological contexts, the preservation of these rachii and their discovery are mainly dependent on whether they became carbonized. As in the case of *Sorghum bicolor*, there are two main factors that determine whether this component is an archaeologically strong isolating variable: the location of the processing and the likelihood of the component being exposed to fire. Threshing and the initial winnowing by wind most often occur in the fields, and thus are less likely to be discovered archaeologically, and their potential for exposure to fire is minimal. Therefore, although the rachii are good indicators of sieving, winnowing by wind and threshing, they do not have a high index of significance as isolating variables.

Rachillas without spikelets can be used as an isolating variable for winnowing by wind byproduct, since they dominate these assemblages (see Table 4-3). They are being processed out predominantly through winnowing by wind, and thus occur in high frequencies in the byproducts of this stage. In contrast, the rachillas with spikelets are not a very discriminating component. However, they do occur in the highest frequencies in the sieving byproduct. Thus if an assemblage is dominated by rachillas without spikelets, then the assemblage is most closely related to winnowing by wind byproduct. On the other hand, if an assemblage has both types of rachillas occurring in moderate amounts it is most closely related to a sieving byproduct. However, such correlations are accurate only when more than one component is used to characterize the assemblages.

Pennisetum typhoides grains with and without husk are important indicators for crop products from all stages, as opposed to crop byproducts (Table 4-3). The key element of this pattern entails the size of the crop grains. The prime grain is predominantly in sieve B, while the tail grain (including the immature grains) is in sieve C. The distinction of *Pennisetum typhoides* grains associated with crop products versus those associated with crop byproducts lies in the sieve size of the respective grains. Those crop byproducts that have crop grains in dominant frequencies, have predominantly sieve C grains, while the crop products, particularly in the later stages, have more sieve B grains. Therefore, grain size measurements are critical in determining the nature of an assemblage dominated by *Pennisetum typhoides* grains.

In sum, the presence or absence of *Pennisetum typhoides* grains solely cannot be used alone to determine a particular processing stage, but their distribution must be used in conjunction with other components. For example, if the *Pennisetum typhoides* seeds with and without husk constitute the majority of the assemblage, then the assemblage could be interpreted to be comparable to a winnowing by wind 2 product, or a sieving product. This reiterates a key point: individual components cannot be used as the singular isolating variables or indicators.

Weed categories are important distinguishing variables, and in conjunction with other components they provide distinct signatures to differentiate various stages' products and byproducts. One distinctive pattern of the weed categories distribution along the processing pathway is that the majority of the weeds are processed out in the later stages, since the focus of the earlier processing stages is on the larger non-crop grain components. This is important for archaeological modeling, since the later stages' products and byproducts are more likely to be discovered archaeologically.

Typically, the higher/later the processing stage the smaller the size of the components: there are negligible amounts of sieve A components, and in the last few stages there are only sieve B and C components. *Pennisetum typhoides* grain occurs mostly in sieve B and C, and when they occasionally occur in sieve A, they are usually the infected bloated grains. This pattern also occurs with *Sorghum bicolor* and is an expected pattern, since the bigger the size of non-crop grain components (that is, sieve A components) the higher the chances of them being processed out early. The larger the component, the higher its visibility, and therefore the more likely it is to be sorted out by hand. In addition, the ethnographic sieving (a relatively early processing stage) also serves to sort out large undesirable components. In relation to this size discrimination along the pathway, the *Pennisetum typhoides* processing study reveals that weeds (which are typically small) are primarily processed out in the later stages. This is a key pattern for archaeological modeling, since the assemblages from these later stages are typically recovered archaeologically.

An Ethnographic Model for the Processing of Type A Crops (*Sorghum bicolor* and *Pennisetum typhoides*)

As discussed in chapter 3, *Sorghum bicolor* and *Pennisetum typhoides* can be harvested in three different methods: (1) a typical Type I harvest method, (2) a typical harvest II method, and (3) an intermediate method similar to Type II harvest (see glossary). The difference in the three harvesting methods has considerable effect on the compositions of the crop products and byproducts at the various processing stages. Of the three methods, the Type I method of harvest was studied, specifically of *Sorghum bicolor* and *Pennisetum typhoides*. This method is the most representative of the three methods. Therefore by addressing this method the research takes into consideration all possible variations in the crop product and byproduct compositions at different stages due to all three methods of harvest. A crop processing model was developed for *Sorghum bicolor* and *Pennisetum typhoides* harvested in a Type II method (Figures 4-7 and 4-8). However, the patterning of the components for the other two methods of harvest is represented within this same model, since it primarily involves inclusion or exclusion of certain crop components. For example, if a Type A crop is harvested in a Type I method, then only the panicle heads will be included into the processing pathway. Therefore, the components of the different products and byproducts will be restricted only to the crop rachii, rachillas, spikelets, and grains. Such variations in the model when the two Type A crops, *Sorghum bicolor* and *Pennisetum typhoides,* are harvested in Type I method will be discussed later in the chapter.

The patterns distinguished in data analysis and discussed in detail previously are summarized in Table 4-4 for *Sorghum bicolor* and Table 4-5 for *Pennisetum typhoides*. These patterns provide the framework for building ethnographic models for the processing of Type A crops growing with weeds harvested in a Type II method. Table 4-4 combines the data from both the *Sorghum bicolor* plots (S1 and S2) while the data from all the three *Pennisetum typhoides* plots (P1, P2, P3) are combined in Table 4-5. The distributions for each component were collapsed into 6 ranges, which were given qualitative terms, ranges, and symbols (see respective tables for details): absent, extremely rare, rare, moderate, strong, and dominant. The resulting tables are the consolidation and expression of these classifications, and provide the framework to understand the crop processing of Type A crops, such as *Sorghum bicolor* and *Pennisetum typhoides* growing with weeds, when they are processed through the following stages: threshing with cattle, threshing with sticks, winnowing by wind, sieving, winnowing by shaking, pounding, and grinding.

In addition to distribution patterns of crop grain and non-grain components down the processing pathway, there are four independent factors that are instrumental in assessing the probability of the different components entering the archaeological contexts. They include likelihood of being prone to accidental or intentional fires (high, moderate, and low likelihood), storage of crop product (unlikely, likely, and definitely stored), fodder use (immediate or stored), and incorporation into dung as temper or fodder and subsequent use as fuel. All the products and byproducts from the processing stages are ranked for each factor. The ranking is not subjective, but based on ethnographic observations, archaeological data, and postdepositional studies.

The ranking according to the liability to accidental burning/fires is critical because this allows for the investigator to recognize which assemblages are most important for archaeological situations. The degree of exposure of a crop product or byproduct to fire is ranked into three levels: high, moderate, and low. If an assemblage has a high index of being accidentally burned, then it has a higher chance at surviving into the archaeological record. It is important to note, however, that there are some components from assemblages (such as straw) that have a moderate to high likelihood of being burnt but a very low chance at surviving into the archaeological record or being discovered archaeologically. Such a component is not a critical isolating variable despite its moderate to high fire index. The fire exposure index of a crop product or byproduct is based on several factors including location of processing, storage index, and the characteristics of the processing stage itself. An assemblage that is definitely stored in domestic areas has a higher chance of being exposed to accidental and catastrophic fire. In addition, the later stages, which are most often processed in close proximity to cooking areas, have a higher index for fire exposure. For example, winnowing by shaking before grinding often occurs in the front/back yards near cooking areas, and the byproducts are often swept into the hearths, and therefore this assemblage has a high fire index (see Figures 4-7 and 4-8).

The next important factor in assessing the probability, significance, and relevance of the different components entering the archaeological context is crop storage. Crop storage categories include not stored for later processing (NSLP), likely to be stored for later processing (LSLP), and definitely stored for later processing (DSLP). A crop product that is definitely stored for later processing has a higher probability of entering the archaeological record (this is also, of course, dependent on the fire liability index), while a crop product that is not stored for later processing has a low to negligible probability. In general, crop products that are closer to consumption have a lower probability of being stored for later processing, however crop products that are ready for consumption also have a high storage likelihood. This storage (of crop products that are ready for consumption) is significantly different than the storage index being discussed here, since the crop products that are ready for consumption are totally clean while the other crop products are not cleaned and have non-crop grain components.

Whether a crop product is stored for later processing is dependent on the next stage of processing, nature of the material, and other demands on the farmer. For example, crop products are typically not stored for long periods after they are pounded, since the grain has to be separated from the chaff dust through winnowing by shaking. Damage through fungal/bacterial infection to crop products may occur if they are stored in such conditions (after pounding but before winnowing by shaking). Thus, the storage factor is also dependent on whether the crop product can survive storage without being subject to fungal/bacterial infection, and becoming prey to small animals such as rodents.

In general, *Sorghum bicolor* and *Pennisetum typhoides* do not store well over long periods of time. This is due to their dehulled nature (naked grains without a pericarp) and hence they run a relatively high risk of damage through insects or fungus during storage. A survey of storage rates in a non-urban context in India revealed that in most households damage to stored *Sorghum bicolor* and *Pennisetum typhoides* was between 90-93%, while in dramatic contrast storage damage to *Panicum miliare* and *Setaria italica* was 2-3% (Pushpamma 1989). Additionally, storage of crops like *Sorghum bicolor* was significantly lower on average when compared to crops such as *Panicum miliare* and *Setaria italica.*

For *Pennisetum typhoides*, rodents are serious pests, which limit production and storage. *Pennisetum typhoides* grain storage is often inadequate, since it is susceptible to storage pests, molds, and rodents. Thus it is important to note that there are diseases of serious consequence to *Pennisetum*. However, the relatively short growth cycle of the crop and its cultivation in dry regions, often directly following a long, hot, dry season (as is the case in Gujarat), may tend to reduce the incidence of serious limiting pathogens. In

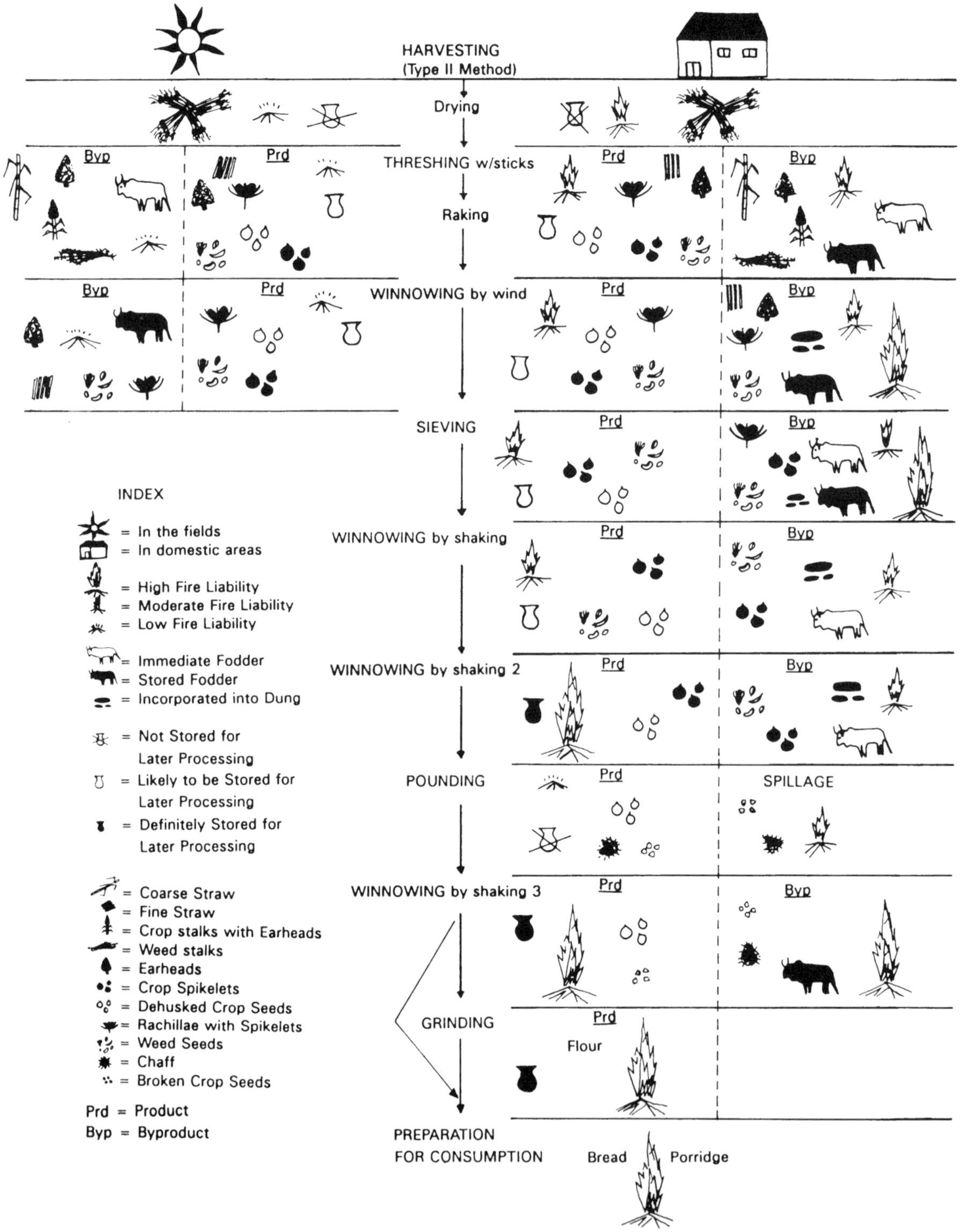

Fig. 4-7. Crop Processing Model for *Sorghum bicolor*.

Table 4-4. Tabulation of *Sorghum bicolor* distributions.

Stages	**Straw**	**Panicles**	**Rachillae**	**Sorghum spikelets**	**Sorghum grain**	**Weeds**	**SHH**	**SHL**	**SFH**	**SFL**	**BHH**	**BHL**	**BFL**
Threshing Product	★	❍	☆	★	★	★★★	❍	☆	★★	★★★	❍	❍	❍
Win(w) Product	❍	❍	★	★★	★★	★★	❍	❍	★★★	★★	–	–	❍
Win(w) Byproduct 1	★★★	☆	☆	☆	❍	★	☆	★★	★	★	–	☆	☆
Win(w) Byproduct 2	★★★	❍	☆	–	–	★	☆	★★	★★	★	–	☆	☆
Sieving Product	❍	–	–	★	★	★★★	❍	❍	★★★	★★	–	–	❍
Sieving Byproduct	☆	❍	★★★	–	☆	☆	❍	★★	★★	★	–	❍	★
Win(s) Product	–	–	–	★★★	★★	☆	★★	–	★★	★	–	–	–
Win(s) Byproduct	☆	–	–	★★	☆	★★★	❍	❍	★★	★★★	–	–	❍

SHH = small headed heavy; SHL = small headed light; SFH = small free heavy; SFL = small free light; BHH = big headed heavy; BHL = big headed light; BFL = big free light

– = Absent
❍ = Extremely Rare
☆ = Rare
★ = Moderate
★★ = Strong
★★★ = Dominant

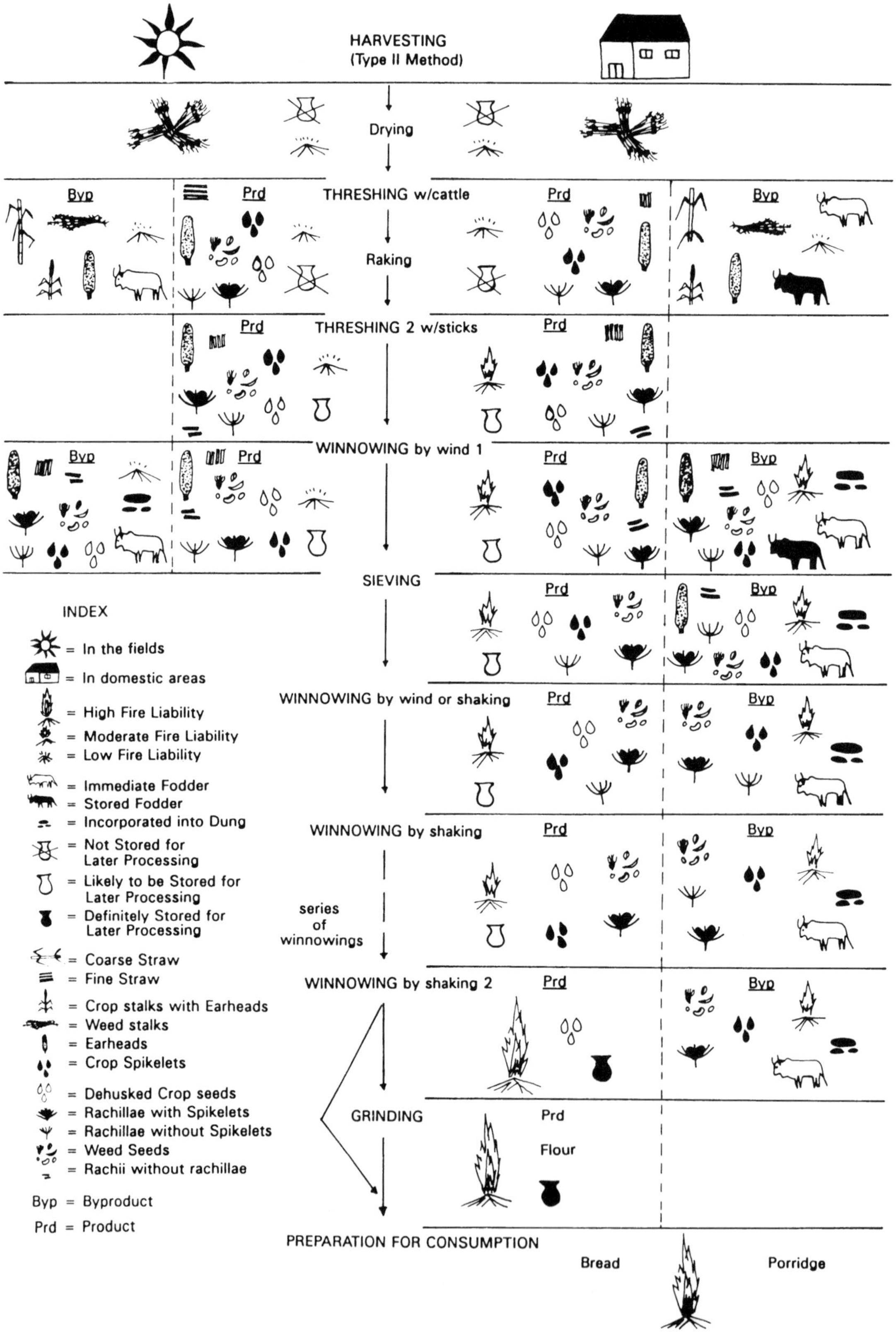

Fig. 4-8. Crop processing model for *Pennisetum typhoides*.

Table 4-5. Tabulation of *Pennisetum typhoides* distributions.

Stages	Straw	Rachii With rachillae	Rachii w/o rachillae	Rachillae with seeds	Rachillae w/o seeds	Penn. spikelet	Penn. Grain	Total Weed	SHH	SFH	SFL	SHL	BHH	BFH	BHL	BFL
Threshing Product 1	❍	❍	–	☆	★	☆	★★	☆	❍	☆	❍	❍	❍	–	–	❍
Threshing Product 2	☆	❍	❍	❍	★★	❍	★	☆	❍	❍	❍	❍	❍	–	❍	❍
Win(w) Product	❍	❍	❍	☆	❍	❍	★★★	☆	❍	☆	☆	❍	❍	–	–	❍
Win(w) Byproduct	❍	❍	❍	❍	★★★	❍	❍	❍	❍	❍	❍	❍	–	–	❍	❍
Sieving Product	–	–	–	❍	❍	☆	★★★	☆	❍	☆	❍	❍	❍	❍	–	❍
Sieving Byproduct	❍	❍	❍	★	★	☆	★	☆	❍	❍	❍	❍	❍	❍	–	❍
Win(w) 2 Product	–	–	–	❍	❍	☆	★★★	☆	❍	☆	❍	❍	–	–	–	–
Win(w) 2 Byproduct	❍	–	–	❍	❍	☆	★★	★	❍	★	☆	❍	–	–	–	❍
Win(s) Byproduct	–	–	–	❍	❍	☆	★★★	☆	❍	☆	☆	–	❍	–	–	–

SHH = small headed heavy; SFH = small free heavy; SFL = small free light; SHL = small headed light; BHH = big headed heavy; BFH = big free heavy; BHL = big headed light; BFL = big free light

– = Absent
❍ = Extremely Rare
☆ = Rare
★ = Moderate
★★ = Strong
★★★ = Dominant

general, recent hybrids are more susceptible to pests and pathogens than the traditional non-hybrids. Loss of *Pennisetum typhoides* grain during storage from rodents alone is estimated to be 25% in India (Rachie and Majmudar 1980). According to Rachie and Majmudar (1980) *Pennisetum typhoides* grain storage in India occurs soon after threshing of a well-dried crop (the drying before threshing ranging from few days to several weeks, depending on weather and other conditions). In addition, good quality *Pennisetum typhoides* heads are sometimes stored for future sowing. Storage is in jute/cloth bags, earthenware, metal, or woven basket-like containers or in pits lined with straw. (It is important to note that *Pennisetum typhoides* grain storage in Africa involves storage of the crop on the head, while in India loose grain is typically stored).

Rachie and Majmudar's (1980) study, however, only deals with larger farms, and it is emphasized that storage patterns vary depending on the size of the farm. Typically, the smaller farms tend to process (clean) the crop into the later processing stages in the immediate weeks following the harvest, while the larger farms tend to store immediately after threshing and before any cleaning. The large farms then market the grain to grain merchants in this condition, while the smaller farms tend to market cleaner crop products. This has significant implications for archaeological interpretations and modeling. For this study, the small farm procedures are used for archaeological modeling of the Harappan Phase in Gujarat, since they are more applicable than the larger farm techniques of modern India.

Use of a crop product or byproduct as fodder is an issue of considerable importance since it relates directly to one of the main themes of this study; the use of millet crops as fodder for animals during the Harappan Phase in Gujarat. Determining whether the crop product or byproduct is used as immediate fodder or as stored fodder is just as important as the question of whether a crop product or byproduct is used as fodder. The index of fodder popularity (with respect to terms of immediate fodder use or stored fodder) helps elucidate the selection of certain products/byproducts for animal feed. Selection for specific stage byproducts (or in rare cases, products) indicates that the selection is based on certain requirements or needs of the animal (for example, fresh fodder is better for lactating animals). Specific selection of a crop byproduct for fodder, and storage of this byproduct for later use as fodder demonstrates the importance of this plant material. Certain crop products may have a high fodder popularity but low storage possibility due to the nature of the plants (i.e., does not store well for long periods). When an assemblage (crop product or byproduct) is used as immediate fodder then the likelihood of this assemblage entering the archaeological record as a distinct collection is minimized. In contrast, when an assemblage is stored for later use as fodder this increases the chances of entering the archaeological record. Of course, this is subsequently dependent on whether the assemblage is carbonized through exposure to fire, and the context in which this occurs (archaeologically discoverable location or not). Whether an assemblage (crop product or byproduct) is stored or used as immediate fodder is dependent on several factors, some of which are measurable. These include: nature of the assemblage (fresh or dry), amount of the assemblage (plenty or little), location of the processing activity (in the fields or in domiciles), and lastly the presence of animals within the household.

When the *Sorghum bicolor* and *Pennisetum typhoides* plants are relatively fresh (green) at the time of harvest and are not well dried, then crop byproducts such as the stalks and straw are not stored for later use as fodder (Figures 4-7, 4-8). Instead they are given as immediate fodder. The storage of wet, green *Sorghum bicolor* and *Pennisetum typhoides* plants runs the risk of a fungal infection on the vegetative parts of the plant which on ingestion by cattle can cause often fatal stomach bloat. Therefore, in the majority of the cases, fresh *Sorghum bicolor* harvests result in the crop byproducts being fed as immediate fodder to animals. This is of significance in places like Africa where the harvest of these millets occurs during the wetter months (Rachie and Majmudar 1980). In contrast, the harvest months in Gujarat (and most of India) are characteristically dry. Hence, the option of drying and then storing fodder is possible in India.

The quantity of material usable for fodder is another critical factor that determines when an assemblage becomes stored fodder or immediate fodder. When there is a considerable amount of material that is well dried, then the assemblage is most often stored for later use as fodder. This decision, of course, is heavily dependent on whether there is other fodder (wild or otherwise) available for the animals at the same time. In general, wild fodder does not store as well as the dry crop harvest residues and therefore when possible, the wild fodders are used first and then crop residues are used. Investment of time and effort in storing fodder is important because it could be used as an indirect indication of the status of animals in the economy. If a crop product is stored for use later as fodder, the implications are that long term planning for animal fodder is a concern, and that the crop byproduct fodder is of value, however small in quantity it might be.

The location of the processing activities is also a contributing factor for when a crop product or byproduct becomes an immediate or stored fodder. These decisions are related to the investment of time and effort into movement of the materials. If there is a lot of material which is dry and can be used as stored fodder, then effort and time are usually invested into transporting this material to the domicile for storage. If it is fresh and in small quantities then it is most often used as immediate fodder. The decision regarding how much fodder is worth storing is dependent on a wide range of factors.

Whether the farmer has animals or not is an obvious determining factor. However, it is not a direct relationship: i.e., if a farmer does not have animals then he does not use the crop harvest residues (crop products and byproducts) as fodder. In most cases, farmers without animals exchange or give fodder to animal owners, and the key factors mentioned previously (nature of the plant material, quantity of material, and location of the processing) determine this exchange.

Incorporation of assemblages and components into dung through tempering or fodder is an important factor to account for the entry of crop byproduct assemblages into the archaeological record through secondary pathways. It is important to be aware of such processes in order not to confuse the interpretation of assemblages found in dung fuel. A further complicating factor is the inability to archaeologically distinguish between crop products incorporated into dung through tempering versus through ingestion as fodder. The verification of the use of crop byproduct as fodder can be done only through the carbon isotope analysis of the animal bones.

Using the above assessing factors and the patterns observed earlier through data analysis, models for the processing of *Sorghum bicolor* and *Pennisetum typhoides* when harvested in Type II method were developed (see Figures 4-7 and 4-8). The models follow processing pathways that include harvesting in a Type II method, threshing with cattle and sticks, winnowings by wind, sieving, winnowings by shaking, pounding, winnowings by shaking, pounding, and grinding before preparation for consumption. There are several variations possible in the processing pathways of *Sorghum bicolor* and *Pennisetum typhoides.* One variation involves the harvest of *Sorghum bicolor* and *Pennisetum typhoides* in a Type I method (i.e., when only the panicle heads are cut, leaving behind the stalks and all the weeds, thus resulting in a clean harvest). A second variation is when byproducts from different processing stages (of both crops) have considerable quantities of desired materials (crop grains). For example, the winnowing by wind byproduct has considerable numbers of unthreshed inflorescence heads and grain. This most often occurs if the wind is too strong or too light. In such instances, the byproduct is not discarded but treated separately through several additional stages (in a parallel pathway) until it can be combined with the rest of the products in the main pathway (Figures 4-3, 4-6). The decision of directing an assemblage/crop byproduct through a more intensive processing sub-pathway is essentially dependent on the quantities of crop grains in the byproduct assemblage. However, the decision-making is based on the experience of the processor, and is also to some degree an unquantifiable aspect of the ethnic group in question. For example, grain retrieval from byproducts was observed to be more prevalent in Gujarat than in Andhra Pradesh. Similarly, such decision-making is also involved in issues related to fodder usage, and the relevance of winnowing by wind versus winnowing by shaking. There are some definite measurable factors in each of these cases. However, decision making of the processor is an important determining factor that is not quantifiable. For example, when an assemblage is winnowed by wind is dependent on how strong the wind is at that particular time. However, it is not a simple matter of delaying the processing until the wind is suitable. It is also a matter of whether the processing can wait, whether it is worthwhile to winnow by shaking instead, and so on. Such decisions are made by the processor based on experience, his/her ethnic group, economic situation, and other factors related to time management.

When *Sorghum bicolor* and *Pennisetum typhoides* are harvested in a Type I method, this involves the harvest of only the (inflorescence) heads. They are cut from the plant and processed for the grain. This results in a 'clean' harvest where there are no weeds, stalks, or straw to be processed out. The resulting processing of these crops involves first the threshing of the heads to separate the grain, and the subsequent processing stages are the same as those presented in the models. The critical difference lies in the compositions of the various crop products and byproducts and these pathways are clearly deductible. Weeds are totally absent, since they were not harvested into the processing pathway initially. Straw is also absent since the *Sorghum bicolor* and *Pennisetum typhoides* stalks were not included in the harvest process.

Therefore in the absence of weeds and straw, the discerning components for a Type I harvest of *Sorghum bicolor* are crop grain with and without husk, *Sorghum bicolor* rachillas with spikelets, and the processed inflorescence heads/rachii. The distribution of these components is presented in the model for the processing of *Sorghum bicolor* when harvested in Type II method. Thus, the *Sorghum bicolor* rachillas with spikelets are processed out significantly through sieving, there is loss of *Sorghum bicolor* grains with husk and without husk during the winnowing by shaking processes, and the inflorescence heads are processed out in the early stages of threshing and winnowing by wind (Table 4-4). The key to the identification of specific crop products is the overall proportions of the different components.

Similarly for *Pennisetum typhoides* harvested in a Type I method, particularly in the absence of weeds, the discerning components are crop grain with and without husk, rachii with and without rachillas, and rachillas with and without spikelets (Table 4-5). All these components are processed out/selected in the same manner as when the crop is harvested in Type II method. It is possible, however, that there might be a reduction in the number of winnowing by shaking stages since there are no weeds to process out.

Of critical importance is that a Type A crop, such as *Sorghum bicolor* and *Pennisetum typhoides*, will never have weeds in crop products at any processing stage when harvested in a Type I method. If an assemblage of Type A crop (*Sorghum bicolor*, *Pennisetum typhoides*) has weeds associated with it then it is certain that the plant was not harvested in a Type I method, and it had to be harvested in a method resembling a Type II harvest method.

There are two important implications of the models for the processing of *Sorghum bicolor* and *Pennisetum typhoides*. First, the composition of the different stage crop products is significantly different and this is direct result of the processing activity. For example, *Sorghum bicolor* rachillas with spikelets occur in significantly dominant frequencies only in the sieving byproduct. Therefore when an assemblage (archaeological or ethnographic) is dominated by them, this is a strong indicator that there is a very high probability that it is related to a similar process (i.e. sieving) byproduct (Table 4-4, Figure 4-7). When seeds (*Sorghum* or *Pennisetum*) dominate an assemblage, for example if the relative proportion of seeds is much higher than other components, then the assemblage could be interpreted as representing a crop product as opposed to a byproduct (Tables 4-4 and 4-5). Similarly *Pennisetum typhoides* rachillas without spikelets can be used as a discriminating variable for the crop's winnowing by wind byproduct since they dominate this assemblage significantly (Table 4-5, Figure 4-8).

Second, analysis of the *Sorghum bicolor* and *Pennisetum typhoides* processing pathways reveals that the larger undesirable components are processed out in the earlier stages and the smaller undesirable components (including the weeds) are processed out in the later stages. For example, the large *Sorghum bicolor* inflorescence heads are definite indicators of lower processing stages such as threshing and winnowing by wind. They, however, cannot be used as the exclusive isolating variables for archaeological contexts since they are quite unlikely to preserve into the archaeological record (Table 4-4, Figure 4-7). Additionally, the higher the processing stage, the fewer sieve A components in the crop products and byproducts in the *Sorghum bicolor* processing. In the last few stages (such as winnowing by shaking 1, winnowing by shaking 2, pounding and winnowing by shaking 3), the crop products and byproducts have only sieve B and sieve C components. Similar patterns were also noted for the *Pennisetum typhoides* processing, where the rachii were clear indicators of the lower processing stages, and they are predominantly processed out through sieving (Table 4-5, Figure 4-8).

Type B Crops: Opportunistic Cultivation of *Panicum miliare*

'Opportunistic' Cultivation of two stretches of a Type B crop, *Panicum miliare,* along two sides of a river was studied in Andhra Pradesh, South India (see Chapter 3, Figure 3-20, 3-21). The study plots from the riverbank on the side of the village Rapanpalli are termed ROC plots (Rapanpalli 'Opportunistic' Cultivation plots), while the study plots on the banks of the town are called SOC plots (Sironcha 'Opportunistic' Cultivation plots). A total of fives study units, approximately 25 m^2 in area, was laid out in different parts of the two fields; they include ROC 1 (furthest away from the water), ROC 2 (closest to water), ROC 3 (midway and near a path), SOC 1 (furthest away from water and closest to bank), and finally SOC 2 (closest to water).

Panicum miliare

Panicum miliare is a Type B crop with a morphology that is distinct from such Type A crops as *Sorghum bicolor* and *Pennisetum typhoides*. It bears the grains at the top of a stem (culm) in the form of a spreading panicle (see Figure 3-21). The panicle consists of a rachis with rachillas at each node. The length of the rachillas and the internodes may vary considerably. The rachillas terminate in numerous spikelets and each spikelet consists of four thin membranous glumes enclosing two florets, of which only one is fertile (Aiyer 1982:91). The fertile floret contains the rounded caryopsis in a shiny lemma and palea (Renfrew 1973). During the processing, the products and byproducts from different stages contain different elements of the plant including: stalks, inflorescence heads (panicles), grain with husk (spikelets), grain without husk, chaff/glumes, straw, and rachillas (Figure 4-9).

As in the case of *Sorghum bicolor* and *Pennisetum typhoides*, the data collection and analysis was a two-step process. During the first stage of analysis a subsample was first examined and the different categories were formed by sorting through the material and categorizing each plant part into the relevant botanical classification.

The second stage of analysis entailed the detailed analysis of the rachillas. This was done because, in the absence of weeds, it was evident that the spikelets were the most distinct and potentially distinguishing category. Weeds have been used as indicators of crop processing stages by Jones (1987), however, the absence of weeds in 'Opportunistic' cultivation necessitates that one rely on crop parts for indications of crop processing stages (Hillman 1985; Reddy 1991c).

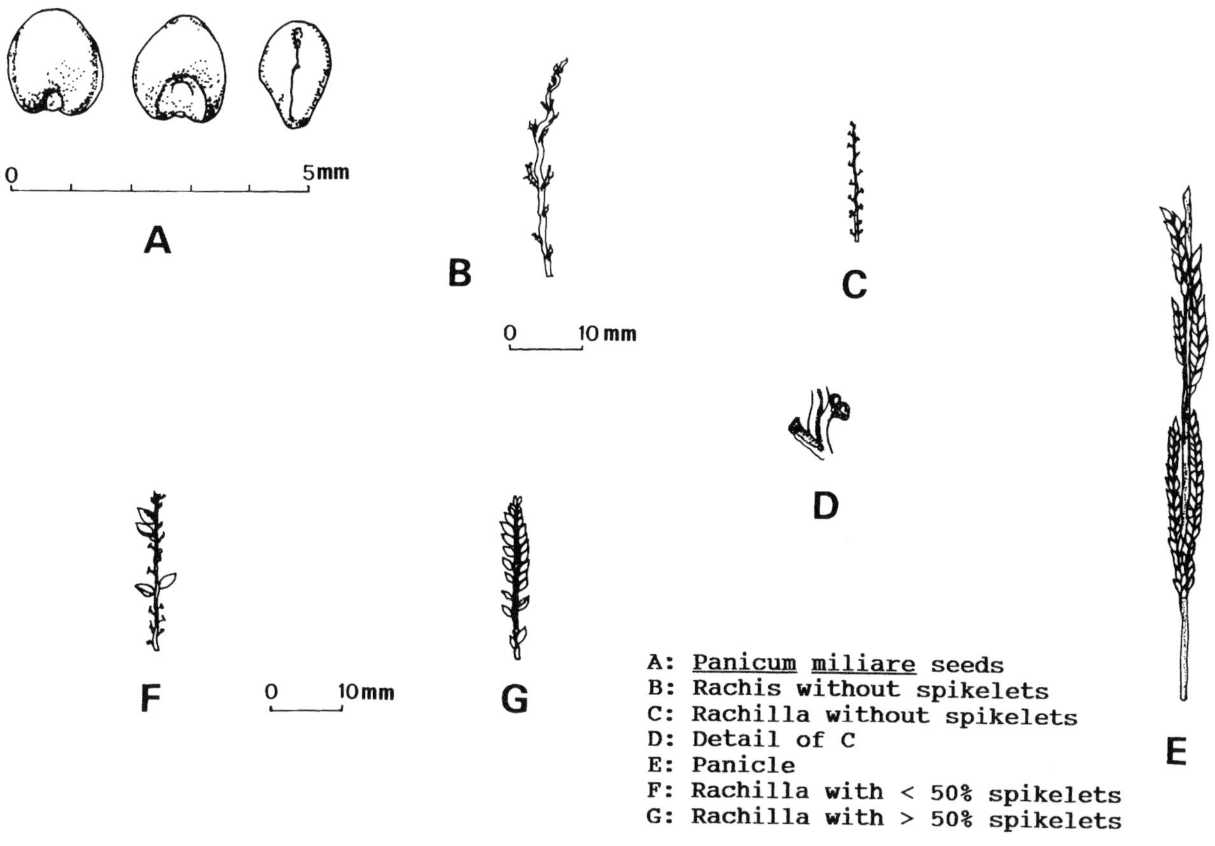

Fig. 4-9. *Panicum miliare* plant showing major components.

The rachillas were categorized as those with more than 50% spikelets, with less than 50% spikelets, and without spikelets (Figures 4-9, 4-10). The hypothesis behind this categorization is that the number of spikelet bearing rachillas will increase in the latter stages while the non-spikelet bearing rachillas will be removed earlier in the process. This procedure would facilitate more recovery of seeds from the spikelet bearing rachillas in the subsequent processing stages and lower the amount of grain lost.

General Patterns

During the initial analysis of *Panicum miliare* samples several general observations about the composition of samples from the various processing stages are noted. First, the higher the processing stage (closer to consumption product) the fewer the categories in the composition analysis. For example there are fewer non-grain components, such as straw, panicle heads and chaff, in the higher processing stages of winnowing II and pounding product. This reduction in the number of categories corresponds to the fundamental principle behind crop processing: to obtain a cleaned crop product, and to do so, the stages closer to the consumption stage have to be cleaner as far as non-grain components are concerned. In addition, the processing activity becomes more coercive and focused on the intense separation of the crop grain from the non-desirables as the stage gets higher.

The second observation is that categories such as straw and panicle heads are present in significant numbers only in the initial stages such as threshing by-product, threshing product, winnowing I by-product, and winnowing I product. The likelihood of their survival is minimal but their use as fodder is high. The disposal of these categories occurs most frequently in fields (locales away from the home base/settlement), which are very unlikely to be discovered. In addition, these categories would be preserved only in the likelihood of accidental burning but despite this they probably would not survive the burning due to their fragile nature. Straw would decompose and be undistinguishable, while panicle heads might survive the burning. It is important to note, however, that these two categories play a significant role as fodder. They are given to cattle in the field, and it is possible that the panicle heads or some parts of them could be found in the dung from these cattle. This

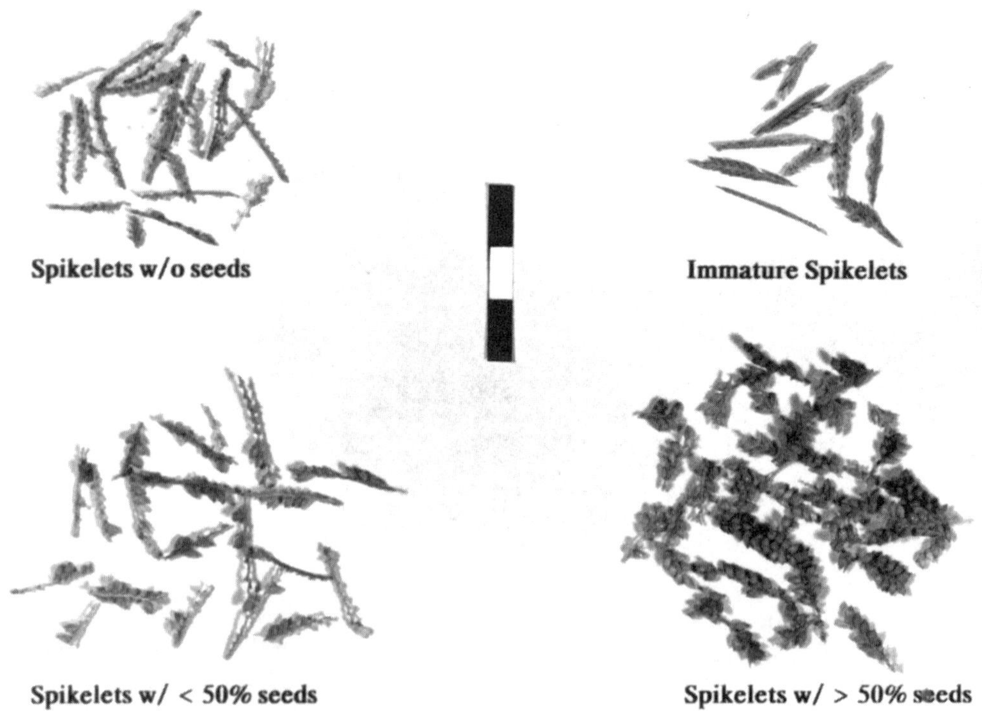

Fig. 4-10. *Panicum miliare* spikelets.

dung could in some form become part of the home base (e.g. dung plastered floors, dung fuels, etc.). Therefore, it is important to examine the characteristics of these categories, even if their direct discovery is improbable.

The third observation is that size of components is an important factor that determines their occurrence in a processing stage. For example, sieve A components occur significantly only in lower processing stages and are absent in the higher processing stages. This is logically predictable since the larger non-grain material is processed out early, while the smaller non-desirables will be processed out later.

Fourth, in the absence of weeds, rachillas are the determining components for identification of processing stages for 'Opportunistic' Cultivation. The total rachillas (those without seeds and those with less than 50% spikelets) decrease as the stage number increases, whereas the rachillas with more than 50% spikelets increase as the stage number increases. This pattern suggests that the rachillas with more than 50% spikelets are not being filtered out initially in order to facilitate the recovery of seeds from them in the subsequent coercive processing stages. This procedure decreases the amount of grain lost during the crop processing.

Detailed Data Analysis

After the samples were sorted and categorization completed, the data were analyzed for patterns relating to the different processing activities and to isolate categories indicative of a particular processing activity. First, the raw counts were standardized (see the discussion on standardization for Type A crops). Since there was no significant difference in the compositions among the three plots from ROC (ROC 1, ROC 2 and ROC 3), they were combined into one set. Similarly the SOC plots (SOC 1 and SOC 2) were combined. There are some differences between the ROC and SOC samples which will be discussed. The results of the analysis and pattern searching are discussed in the following order: Straw, Panicle Inflorescence, Chaff, Grain with husk, Grain without husk, and finally the different types of rachillas.

Straw and Panicles

Straw is present in significant quantities in only two stages: threshing product and winnowing I byproduct (Table 4-6). Of these two stages, the winnowing I byproduct has straw in proportionally higher frequencies as compared to other components. Although straw is well represented in both stages, its distribution varies between sieve sizes. Sieve C straw is absent in winnowing I byproduct, but present in significant quantities in threshing product. This difference could be attributed to the winnowing process, which separates the lighter material from the heavier material. In this case the straw was blown off beyond the sample collection area, therefore its notable absence. Although sieve A straw occurs in similar frequencies in both stages, there is significantly more sieve B straw in winnowing I byproduct. This is due to the breakage of sieve A straw during the winnowing process, especially since the ROC crop was well matured and very dry.

The two assemblages (threshing product and winnowing I byproduct) are also characterized by the presence, albeit in minimal quantities, of panicle inflorescences. The panicles dominate the threshing byproduct, which was not collected for study since the probability of finding and distinguishing it archaeologically is non-existent in South Asia. In general, panicles are significantly more infrequent than in ROC, and occur only in the threshing product. This low frequency of panicles is because the SOC plants were not dry at harvest time, but were partially green. Thus, the panicles were not brittle and did not to break off the plants during threshing (which was done through the rubbing of the panicles between feet). The combination of a partially green crop and threshing by rubbing between feet (which generates less breakage of the inflorescence heads from the rest of the plant), results in significantly lower frequencies of straw and panicle heads in the resulting assemblages. In addition, the absence of sieve C straw in any of the SOC samples supports the argument of less breakage.

The presence of straw and panicles only in the initial lower stages of processing, such as threshing product and winnowing I byproduct, is explained by the basic goal of the processing which is to separate the non-crop grain components from the crop grain components. The logical first step is to separate the biggest and the most obvious components first, and straw and crop inflorescences fall into this category. Therefore, they occur in dominant quantities only in the initial processing stages. In addition, the SOC samples illustrate the close correlation of the state of crop at harvest (in this case well dried or partially green) and the threshing type, and the effects on the resulting compositions.

Chaff

The chaff component occurs in all the stages except the winnowing I and II products (Table 4-6). The chaff always occurs in the sieve C. The chaff from the SOC samples is less dry, and had a greener and softer texture. This chaff is more attractive for cattle fodder than the drier ROC chaff, mainly because it is more palatable and digestible. Chaff is a component that cannot be used as a definite indicator of a processing stage, although its absence might be used as a positive sign of the winnowing product assemblage. However, chaff is very delicate, cannot survive high temperatures, and due to its fragility the likelihood of it surviving into the archaeological record is very slim. If one is fortunate enough to find chaff archaeologically, it would be remarkable, and will definitely indicate the associated material to be an uncleaned assemblage. Just because chaff has low potential for archaeobotanists, should not diminish its status as a significant component in crop processing. Chaff is an attractive fodder, and also an attractive fuel for particular activities such as firing of stones to make beads. It is also used as temper in making dung fuel cakes, and used as temper in plastering floors, walls and roofs. Its archaeological discovery might seem unlikely, but its various uses make chaff a valuable component to consider in the issue of fodder, and the role of millets and their byproducts in a subsistence economy.

Crop grains with husk

This component occurs in all the stages products and byproducts, albeit only in sieve C and in highly differing quantities. Crop grains with husk occur in significant numbers in winnowing I product and threshing product. They occur in the other stages but in highly varied frequencies.

Crop grains without husk

Crop grains without husk (dehusked) are very significant for archaeological modeling. They occur at very specific points in the processing: after

Table 4-6. *Panicum miliare*. Crop Processing Stages Composition.

Counts for *Panicum miliare* per 500 g for each stage

ROC Plots	**Threshing Product**	**Win(s) Product**	**Win(s) Byproduct**	**Win(s)2 Product**	**Win(s)2 Byproduct**
Straw Vol. (ml)	3671	0	4638	0	0
Chaff Vol. (ml)	804	0	2500	0	1602.6
Panicles	2	0	3	0	0
Total Rachillae	1974	340	7380	0	0
Rachillae w/o seeds	932	92	4400	0	0
Rachillae w<50% seeds	550	112	1718	0	0
Rachillae w>50% seeds	492	135	1262	0	0
Grain w/husk (spikelets)	166461	210554	510	56.1	1871.8
Grain w/o husk	0	0	0	552293	1341.6

SOC Plots	**Threshing Product**	**Win(s) Product**	**Win(s) Byproduct**	**Win(s)2 Product**	**Win(s)2 Byproduct**
Straw Vol. (ml)	71	0	1563	0	0
Chaff Vol. (ml)	507	0	3750	0	1776
Panicles	12	0	0	0	0
Total Rachillae	124	12	1433	0	0
Rachillae w/o seeds	68	2	850	0	0
Rachillae w<50% seeds	33	0	246	0	0
Rachillae w>50% seeds	24	10	338	0	0
Grain w/husk (spikelets)	190541	223081	2654	53	1364
Grain w/o husk	0	0	0	626298	1064

the dehusking process which done by pounding in 'Opportunistic' Cultivation. The presence of dehusked crop grains indicates that the assemblage is from a higher processing stage, since these grains do not occur in the earlier processing stages. They occur in significant numbers in winnowing(s) II product, although there are some present in winnowing(s) II byproduct (Table 4-6). The presence of dehusked grains reveals two insights in human activities; first, the grain has been processed for human food and is ready for consumption either as grain or for making flour. The second deduction is that it is not being stored for long, since long term storage of *Panicum miliare* is done in a husked condition. The husk gives the grain protection against fungal attack and moisture, and in a dehusked condition the longevity is problematic.

Rachillas

Distinct patterns in the distribution of all three types of rachillas combined were observed across the processing stages resulting in several conclusive observations. There is a distinct pattern of distribution of all the three types of rachillas between stages. The rachillas occur in significantly higher ratios in the winnowing I byproduct samples while the winnowing I product has the lowest ratio (Table 4-6). This pattern was constant for all three types of rachillas (without spikelets, with <50% spikelets, or with >50% spikelets). The samples of winnowing I byproduct are characterized by significantly high frequencies of rachillas of all three types (as compared to other components), while the samples of winnowing I product, winnowing II product and byproduct, and pounding I product do not have rachillas dominating their assemblages. This supports the hypothesis that as the processing stage increases, the number of components and the non-grain components decrease. Thus, an archaeological sample of crop Type B with a high frequency of rachillas can be interpreted as related to a winnowing I byproduct.

To further examine the distribution of rachillas, pattern searching between rachillas in different sieve sizes was done across all the processing stages. Winnowing I byproduct has the highest percentage of rachillas in each sieve size, which also means they have a higher probability of getting into the archaeological record. This confirms that there is no sieve size relation factor limiting the dominant occurrence of rachillas in this stage byproduct. Their discovery likelihood is not very high, however, since most often the winnowing process is done away from the home bases. If the winnowing process is done at home bases, then the high frequency of rachillas make winnowing I byproduct a distinct assemblage.

Winnowing I product has the lowest frequency of rachillas, of which most are sieve B size. This pattern implies efficient cleaning of sieve A and sieve C rachillas, but poor cleaning of sieve B rachillas. The reason for this situation is that sieve A rachillas are large enough to be picked out by hand sorting while sieve C ones are small and light, and fly out easily. Thus, sieve B rachillas are more likely to be left behind in the winnowing I product. This distribution pattern is noteworthy and important because it contradicts the logical assumption that the smaller the components (as seen in their sieve size distribution) the less efficient the cleaning process.

The winnowing I byproducts from the two study areas (ROC and SOC) are distinct. The ROC winnowing I byproduct has more sieve B than sieve A or C rachillas, while SOC winnowing I byproduct has more sieve A rachillas followed by sieve B and lastly by sieve C rachillas. For ROC, the pattern is attributed to the brittle nature of the rachillas that are broken in the winnowing process, i.e., sieve A rachillas of threshing product are getting broken up into the sieve B size rachillas in the processing, creating more sieve B rachillas in winnowing I byproduct. In the SOC samples, the rachillas are not brittle but instead are malleable since the crop was partially mature and greener at harvest. Therefore, in the SOC samples there is no breakage of the rachillas during the winnowing process. This is significant because it is an important isolating characteristic to identify the nature of the crop at harvest time, and this could shed light on other activities such as the type of threshing that was needed.

One important observation is that there are significantly fewer sieve C rachillas in the SOC samples, which seem to occur only in the winnowing I byproduct. This pattern may be partly due to the type of threshing done to the SOC samples (threshing by rubbing the panicles between feet). This resulted in less breakage of the rachillas, and just dislodged the seeds from the rachillas into independent spikelets. This patterning has significant implications for interpreting an assemblage with respect to the nature of the original crop at harvest and type of threshing needed.

Rachillas do not occur in winnowing II products or byproducts, and pounding I products or byproducts (Table 4-6). The rachillas occur significantly only in the initial stages of processing. Usually intensive cleaning occurs during the winnowing I process. However, if another cleaning is needed, the process is repeated again prior to pounding and usually through winnowing by shaking. This is not a universal pathway, but it is highly likely that the rachillas noted in winnowing I product will be processed out through winnowing by shaking (different from winnowing II) before pounding I. The disposal of these rachillas will be at the home base, usually in the immediate surroundings of the consumption and domestic activity areas. Therefore, these rachillas have a high fire liability and are important for archaeological modeling.

It is important to emphasize that the processing stages being considered are not necessarily the absolute and only way the processing is done at any given time. The order of the processing is absolute, but how many times a single process is done (without it being considered a separate process) is very varied from household to household and from region to region. For example, the winnowing I process when done by shaking can occur as a single event or a series of events. Winnowing I is thus done in a graded manner, initially in a less effective manner and then in an intensive manner. This variability does not invalidate the results discussed above, but one must consider that there might be more than one winnowing I byproduct for a given crop being processed. Each byproduct and product, however, would have the same broad set of characteristics already discussed.

When the distribution of the three rachillas types (without spikelets, with <50%, or with >50% spikelets) between specific sieve sizes was analyzed, a series of patterns was discernible. Similar to the total rachillas patterns, winnowing I byproduct has the highest frequency of rachillas without spikelets (81%), with threshing product (17%) and winnowing I product (2%) in lower frequencies. In winnowing I byproduct rachillas without spikelets occur in higher frequencies in sieve B than in sieve A, and this pattern is consistent with the argument made earlier regarding breakage of rachillas during the winnowing process. There is efficient cleaning of rachillas lacking spikelets during the winnowing process, with winnowing I product having primarily sieve C rachillas, fewer sieve B rachillas, and no sieve A rachillas. This trend supports the logical inference that as the sieve size gets smaller the cleaning becomes less efficient. The presence of rachillas without spikelets in the winnowing product (albeit in low frequencies) also reveals that another cleaning was planned before the pounding process.

The rachillas with < 50% spikelets also occur in highest frequencies in the winnowing I byproduct (72%). The winnowing I product has only sieve B and sieve C components, sieve B having higher frequency than sieve C.

Winnowing I byproduct also has the highest frequency of rachillas with >50% spikelets (67%). Winnowing I product has only sieve B sized rachillas with >50% spikelets, while threshing product has them only in sieve A and sieve B. This differential stage distribution between sieve sizes is interesting because it reflects both the effects of the processing activity and the morphology of the rachillas. The sieve C rachillas with >50% spikelets are absent in the winnowing I product because during the processing activity sieve C rachillas of all the three types were being processed (broken down) into rachillas without spikelets and free grain. This was done by separating the husked grain from the rachillas through manual rubbing of the rachillas against the winnowing basket. The processing ensured a higher retrieval of grain from the non-grain components. This processing creates a low frequency or absence of sieve C rachillas with > 50% and with <50% spikelets.

Summary of Panicum miliare

There are several conclusive statements that can be made through the data analysis and pattern searching of the 'Opportunistic' Cultivation samples. First, winnowing I product has no any sieve A rachillas and sieve C rachillas occur in very low frequencies. Sieve B rachillas occur in the highest frequencies. Of the three types of rachillas in sieve B, those with >50% spikelets occur in the highest frequencies, and would be most indicative of winnowing I product when they occur associated with husked grain, and an absence of straw, panicle inflorescences, or chaff. This pattern is due to several factors. First, sieve C rachillas in general are present in lower frequencies (especially in the SOC samples) due to the non-brittle nature of the rachillas. Second, sieve A rachillas are easier to clean out by hand selection since they are more visible. And lastly the rachillas with > 50% spikelets are difficult to clean because they have considerable grain, and most often the processor tends to keep these rachillas until the ultimate processing stage in order to retrieve more of these grains. This is in stark contrast to the handling of the rachillas without any spikelets, which are processed out almost instantly if they are visible.

In winnowing I byproduct the rachillas occur in very high frequencies, with rachillas without spikelets dominating. Certainly, the winnowing process is efficient in cleaning out the rachillas. The process of cleaning along with the effort to separate crop grains from the rachillas creates a differential treatment of sieve C rachillas during processing. Sieve B rachillas with >50% and <50% spikelets are not treated in the same manner and this significant differential treatment can be used to distinguish the processing stage. When sieve B rachillas with spikelets occur in dominating frequencies then the sample could be related to a winnowing I product. In contrast, a winnowing I byproduct sample has more sieve B rachillas without spikelets.

Conclusions of Panicum miliare *analysis*

Using the rachillas data distributions it is possible to define certain patterns which are indicative of assemblages pertaining to threshing product, winnowing I byproduct, and winnowing I product. The winnowing II product and byproduct, on the other hand, can be distinguished on the basis of the distinctive ratios of crop grains with husk to those without husk, and chaff.

Winnowing I product and threshing product are two assemblages that have a higher likelihood of being discovered archaeologically, particularly as compared to winnowing I byproduct. Winnowing I product can be distinguished from threshing product based on the following distinctions:

1. Threshing product has sieve A rachillas and rachillas with <50% spikelets while both are absent in winnowing I product.

2. Winnowing I product has a significantly lower frequency of rachillas without spikelets as compared to threshing product.

3. Most importantly, winnowing I product has a greater crop grain to rachillas ratio than threshing product. In other words, there are fewer rachillas to crop grain in the former assemblage, while the number of rachillas to grain in threshing product is significantly higher.

Even though winnowing I byproduct has a lower likelihood of being discovered archaeologically, it is important to present its distinguishing characteristics as they pertain to rachillas. This assemblage has sieve C rachillas, which are absent or occur in significantly lower frequencies in the other assemblages. The assemblage also has a high percentage of rachillas without spikelets. Most importantly, the ratio of crop grain to rachillas is very low, i.e., there are a lot of rachillas for each grain, and this characteristic sets this assemblage apart from threshing product and winnowing I product.

Differences between ROC and SOC

The main difference between the samples from the two areas, the ROC study plots and the SOC study plots, is the maturity level of the crop at the time of harvest. In addition, the two study plots were threshed using two differing methods. These two differences have ramifications for the compositions of the crop products at various stages.

There is more loss of crop grain in the ROC samples than the SOC samples. This is due to the brittle, well matured, dried panicle in the ROC samples. The maturation condition of the crop plants at the harvest has direct effects on the yields. The yields from the two study area plots differ in that the SOC plots have a better yield than the ROC plots.

In general, the SOC sample processing was more labor intensive, but with a corresponding lower rate of grain loss. This is a direct result of the maturity of the crop plant at the time of harvest. The byproduct from the threshing process of the SOC samples is more suitable for cattle fodder since it is greener and has less celluloid, and thus is easily digestible. The ROC threshing byproduct is totally dried, harder to digest, and less attractive as fodder, although it was fed to animals. Certainly, the greener SOC threshing byproducts are more popular than the ROC threshing byproducts for fodder.

The general cleaning process is easier for the SOC samples, essentially because there are significantly fewer rachillas in the crop product that need to be processed out. Additionally there are fewer inflorescence heads to be cleaned for the SOC samples as compared to the ROC samples. This is a direct result of the maturity of the crop during harvest and the threshing method used. Threshing by rubbing of the seed bearing panicle inflorescences between the processor's feet can be done only when the plant is not too mature, and this method of threshing yields less breakage of the rachillas. The rubbing dislodges the seed from the rachillas rather than breaking the rachillas from the inflorescence heads as it occurs when threshing with a stick. The latter method of threshing is done when the crop plants are well-matured. Thus the combination of the level of maturity (high maturity) and threshing with sticks contributes to there being higher numbers of rachillas in the crop products for the ROC samples. This requires more cleaning processing, particularly in the latter stages.

Since the SOC grains are fresher, roasting is a useful technique for the separation of the husk/glumes from the grain. Thus, there is a higher likelihood of the SOC crop grain being subject to accidental fires because it is roasted before the pounding stage. Spillage during roasting (a very likely scenario) would carbonize the grain and increase its chance of survival into the archaeological record. Crop grains from the ROC samples do not need to be roasted before pounding, since they are well dried. Thus, there are fewer chances of the crop grains from the ROC samples being subject to accidental fires.

If there were weeds associated with these samples, their inclusion into the crop products would vary depending on whether the crop plants were harvested ripe or green, and then how they were threshed. For example, if the crop plant were harvested green and threshed by rubbing the inflorescence panicles between feet, then only the weeds that were tall enough to reach the height of the inflorescence heads at the time of harvest would be included in the threshing. Thus, only a limited number of a specific type of weeds (erect weeds) would enter the processing pathway. In contrast, if a crop plant was harvested in a well matured condition and threshed by using sticks, then all the weeds growing in the plot would be included into the threshing (since the harvesting is a Type II harvest). Thus, the range of weeds being included into the processing pathway would be greater. This is a very important issue to take into consideration when modeling for archaeological interpretations, and one way to avoid the possibility misinterpreting the presence or absence of weeds is to examine the nature of the weed plant morphology (i.e., if they are prostrate type, erect, both, etc.).

An Ethnographic Model for the Processing of Type B Crops without Weeds (*Panicum miliare*)

Patterns in the distribution of crop components between processing stages discerned through data analysis are summarized in Table 4-7. These results form the foundations of an Ethnographic Model for the Processing of Type B crops without weeds (Figure 4-11), exemplified by *Panicum miliare* when it is grown in conditions where there are no associated weeds, and it is processed through threshing with sticks and rubbing of feet, winnowing by shaking, and pounding. The table was developed by combining the standardized counts from the two different field locations (ROC and SOC), and then their distributions were collapsed into five ranges which were given qualitative terms, ranges, and symbols (the ranges vary depending on the data being tabulated for modeling).

As discussed for the Type A Model, four additional factors instrumental in assessing the probability of an assemblage entering the archaeological record have to be considered. The four factors are: likely to be prone to accidental and intentional fires (from degree of very likely to low likelihood); storage of crop product (likely or definitely); fodder popularity (immediate or stored); and inclusion or not into animal dung as temper.

One important distinction of *Panicum miliare* from the storage of Type A crops is that in the case of *Panicum miliare* the risks of fungal and bacterial infection increase as the crop product gets closer to consumption stage, since it is dehusked in these later stages and is vulnerable to such infections. The husk of the grain provides a protective covering, and in its absence, the grain is exposed to infections. In most cases therefore, the transport of grain (particularly to long distance markets) occurs when the grain is still husked. Thus, transport has to occur right after or right before the roasting stage. This pattern has considerable applicability to archaeological interpretation since it provides insights into the ramifications of processing stages with respect to economic processes such as the trade of crop grain.

A model for the processing of *Panicum miliare* (Type B crop without weeds) is presented in Figure 4-11 (see also Table 4-7). The processing pathway is very straight forward, consisting of harvesting, drying, raking, winnowing 1 by shaking, roasting, pounding, winnowing 2 by shaking and grinding, followed by consumption preparations. Several implications are summarized from the model. First, there are specific variables that distinguish the different stages' products and byproducts. For example crop grains with husk occur in significant frequencies in winnowing I product and threshing product, while crop grains without husk occur in significant frequencies only in winnowing II product. Thus ethnographic or archaeological samples showing similar patterns could be interpreted as being most likely related to the particular stage products. Similarly, rachillas in general (and particularly those without spikelets) occur in significantly higher ratios in the winnowing I byproduct, and are strong indicators of this assemblage.

Second, in the ethnographic situation, weeds were notably absent, due to the nature of the soils, and therefore the processing model emphasizes other components such as rachillas as indicators of processing stages' crop products and byproducts. Rachillas are distributed in a patterned manner

Table 4-7a. Tabulation of *Panicum miliare* distributions.

Stages	Straw	Chaff	Panicles	Rachillae w/o	Rachillae w/<50%	Rachillae w>50%	Rachillae w/husk	Spikelets w/o husk	Grain
Threshing Product	☆	☆	☆	☆	☆	☆	☆	★★★	–
Winnowing by shaking 1 Product	–	–	–	☆	☆	☆	☆	★★★	–
Winnowing by shaking 1 Byproduct	★★	★★	☆	★★	★★	★	★	★	–
Winnowing by shaking 2 Product	–	–	–	–	–	–	–	☆	★★★
Winnowing by shaking 2 Byproduct	–	–	–	–	–	–	–	★★	★★

†– = Absent ☆ = Rare ★ = Moderate ★★ = Strong ★★★ = Dominant

Table 4-7b. Tabulation of *Panicum miliare* rachilla distributions by sieve size.

Stages	Total Rachillae			Rachillae w/o seeds			Rachillae w/<50% seeds			Rachillae w>50% seeds		
	A	B	C	A	B	C	A	B	C	A	B	C
Threshing Product	★	★	☆	★	★	☆	★	☆	☆	★	☆	–
Winnowing by shaking 1 Product	–	☆	☆	–	☆	☆	–	☆	☆	–	☆	–
Winnowing by shaking 1 Byproduct	★★	★★★	★	★★	★★	★	★	★	☆	★	★	☆
Winnowing by shaking 2 Product	–	–	–	–	–	–	–	–	–	–	–	–
Winnowing by shaking 2 Byproduct	–	–	–	–	–	–	–	–	–	–	–	–

A= 4mm B= 2mm C= 0.425mm
†– = Absent ☆ = Rare ★ = Moderate ★★ = Strong ★★★ = Dominant

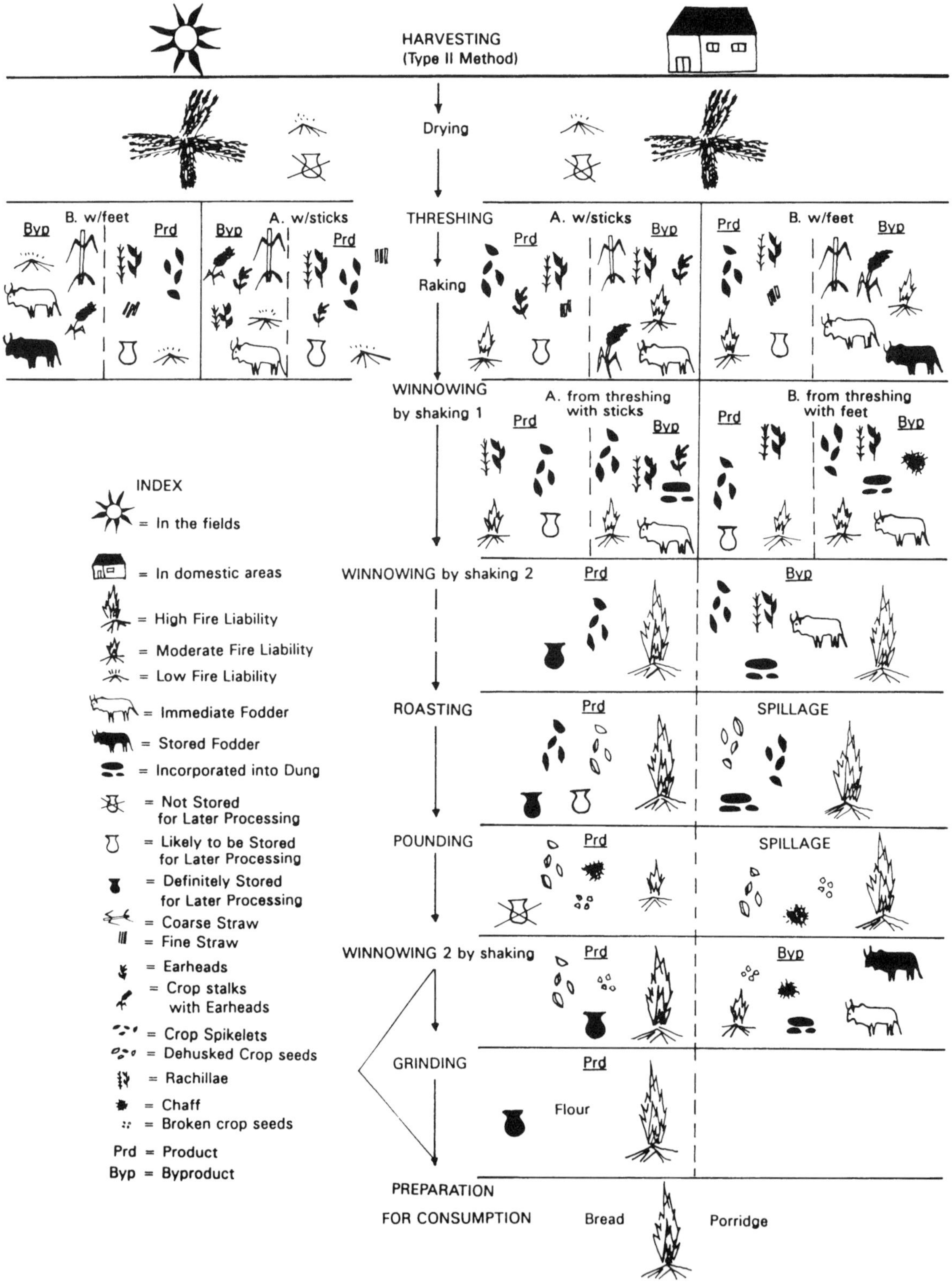

Fig. 4-11. Crop processing model for *Panicum miliare*.

between stages and can be used as definite indicators. Figure 4-11 graphically shows how rachillas fall out in a specific patterned manner in the processing pathway. For example, winnowing I byproduct is characterized by significantly dominating frequencies of rachillas without spikelets, while winnowing I product is characterized by high frequencies of sieve B rachillas with >50% spikelets.

Among the processing stages where rachillas could be potentially distinctive isolating variables, threshing product and winnowing I product are the most likely to survive into the archaeological record and be discovered. This is because winnowing I product is often stored in home bases/ domestic areas, which are most often discovered and excavated. Threshing product is also occasionally stored in domestic areas/ home bases for later processing when other pressing work is present or bad weather occurs. Due to their proximity to domestic areas, their assemblages are more prone to accidental burning in specific contexts and survive into the archaeological record. Given this situation, these two assemblages can be distinguished using the rachillas data. First, the threshing product has sieve A rachillas, while winnowing I product lacks this size rachillas. Second, a higher percentage of sieve B rachillas without spikelets occur in the threshing product sample, while winnowing I product is dominated by sieve B > 50% spikelets rachillas. Most importantly, the ratios of rachillas to husked crop grains are higher for threshing product than winnowing I product. This last pattern is the most distinct feature that can be used to differentiate the two assemblages.

The winnowing byproduct assemblage has a lower probability of being discovered archaeologically because more often the winnowing I process (by wind and shaking) occurs away from the home/ domestic areas. If it does occur in the home/domestic areas, then this assemblage is most often fed to the cattle, and it does not make it into the archaeological record as a distinct assemblage, but some components could survive archaeologically in dung fuel, or in dung plastering material. Thus, it is important to discuss the distinguishing rachillas, which will identify the assemblage. The most important feature is the high ratio of rachillas to husked grain. This pattern of many rachillas and few grains is just the opposite of a threshing product. The rachillas also will be characterized by a predominance of sieve C sized ones, particularly > 50% spikelet rachillas.

Third, no one variable can be used in isolation to define a particular processing stage or stage product or byproduct, but it is a combination of various components in relative proportions that are used as indicators.

Type B Crop Plant with Weeds (*Setaria tomentosa)*

During the study of crop processing of summer monsoonal crops in Gujarat, the processing pathway of a plot of *Setaria tomentosa* was also studied. This was done to compliment the crop processing study of *Panicum miliare*, which was a Type B crop but due to the particular cultivation ('Opportunistic' Cultivation) did not have any associated weeds. The study of one plot of *Setaria tomentosa* provides insight into the crop processing of a Type B plant with weeds associated. For the purpose of this study, *Setaria tomentosa* is treated as a Type B crop plant, since its morphological characteristics are the same. In reality it is a weed associated with disturbed areas and millet fields. In addition, this study can also be used to build models for the cultivation and processing of a wild/undomesticated plant.

Setaria tomentosa

Setaria tomentosa is a small seeded annual grass, which can be often found growing wild in the back gardens of village houses. Its seeds are significantly smaller than those of *Pennisetum typhoides* and *Sorghum bicolor*. The grass is eaten by cattle, and its growth correlates well with the monsoon season. It is considered a drought resistant plant useful for its grain and more importantly it is used as fodder (Figure 4-12).

In the study field, high concentrations of *Setaria tomentosa* occurred along the edges of the *Sorghum* field in a visually distinct, 6 to 7 m wide, and 9 to 10 m long strip (see Figure 3-5). Interviews with the farmer revealed that the strips of *Setaria tomentosa* were not an intentional part of the *Sorghum* field, but were essentially growing as weeds in fallow areas. The few *Sorghum* plants growing in the *Setaria tomentosa* strip were intrusives from last year, or spillage during the sowing of the *Sorghum* field itself and constitute less than 20% of the plot's plants. The *Sorghum* plants in this plot were very thin, i.e., the stalks were very much like Type B crops. Therefore the harvesting was done in the Type II manner. This provided an opportunity to examine how two different crops are processed using the same pathway.

The *Setaria tomentosa* study plot was close to the cattle shed, the well, and adjacent to a pathway in the farmstead. The plot was studied in the manner as those of the other crops addressed in this study. The plot was harvested in a Type II method, with a number of plants gathered at the base and cut with a sickle.

Although *Setaria tomentosa* is a weed for the purpose of processing, the farmers treated it as a crop plant. There were 4 processing stages for *Setaria tomentosa*: threshing with sticks, sieving, winnowing by shaking and pounding. This processing pathway closely mirrors that of *Chenopodium album* (another Type B crop), which is cultivated in the area as both human food and animal fodder. The processing stages and the different methods of crop processing (threshing with stick, sieving and winnowing by shaking) were not any different from those done for *Sorghum* and *Pennisetum,* and are not discussed again for *Setaria tomentosa*.

Detailed Analysis

The samples collected had components of all the three sieve sizes (A, B and C), except winnowing by shaking product and byproduct, which had components of only sieve size B, and C. The different components distinguished during composition analysis of the samples from *Setaria tomentosa* crop processing are: straw, *Setaria tomentosa* rachillas, *Setaria tomentosa* seeds, *Sorghum* rachillas with spikelets, *Sorghum* inflorescence heads, *Sorghum* grain with husk and without husk, and an array of weeds.

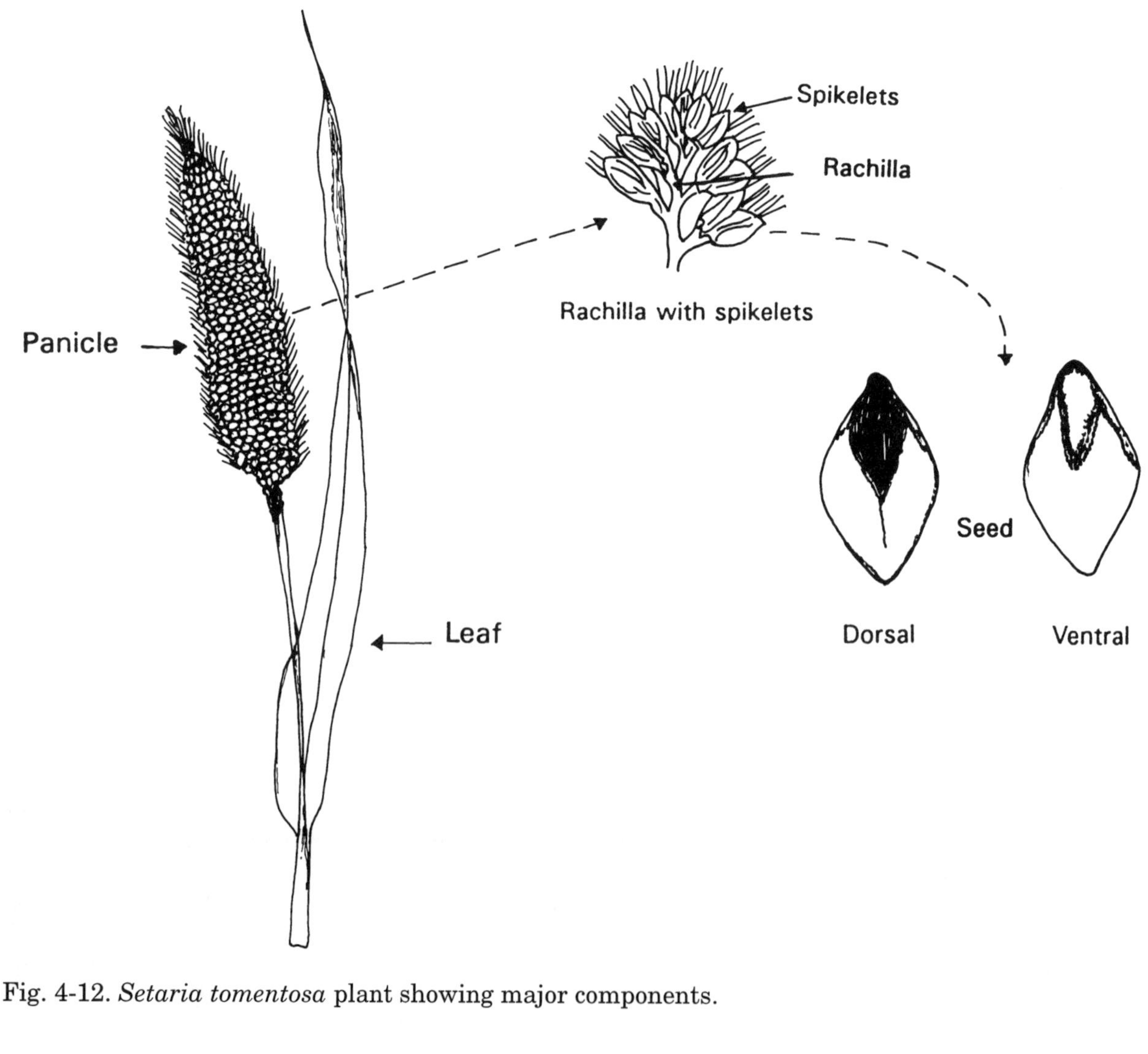

Fig. 4-12. *Setaria tomentosa* plant showing major components.

Straw

Straw occurs in dominant proportions in the sieving byproduct and in strong frequencies in the winnowing by shaking byproduct (Table 4-8). Straw as a component cannot be used as one of the strong indicators of processing, mainly because due to its nature it has low probability of surviving into the archaeological record. Nonetheless it is an important component because of its use as fodder for animals.

Sorghum inflorescence heads

Sorghum inflorescence heads occur significantly in the sieving byproduct, and rarely in sieving product. This component is processed out in the initial stages mainly because of its size (all are sieve A size category), and rarely makes it to a latter stage. The inflorescence heads could be used as definite indicators of initial processing stages if they are found archaeological contexts.

Sorghum *grain with husk*

Sorghum grains with husks occur only in the winnowing by shaking product (Table 4-8). They are not processed out in the sieving or winnowing. There is obvious selection for these grains by the processor despite their low frequencies. They are not selected out because they occur in manageable frequencies (in the processor's perspective) and they can be dehusked manually in the winnowing basket during the winnowing by shaking process.

Sorghum *grain without husk*

Sorghum grain without husk occurs in significant frequencies in three samples: sieving product, and winnowing by shaking product and byproduct (Table 4-8). The high frequencies in sieving product and winnowing by shaking product are expected, since they are being selected. The high frequencies of *Sorghum* grain in the winnowing by shaking byproduct are unexpected, and indicate loss of grain in the winnowing process, suggesting an inefficient process or processor. Detailed examination of the *Sorghum* grain in the winnowing by shaking byproduct reveals that the most of the grains (over 75%) are the tail grains and defective grains (inflected with fungus or decayed). So, in actuality the winnowing process is efficiently removing unsuitable crop products.

Sorghum *rachillas with spikelets*

Sorghum rachillas with spikelets occur in threshing product and sieving byproduct, which demonstrates that the use of an ethnographic sieve efficiently separates them from the crop product sample. Thus, *Sorghum* rachillas with spikelets can used as indicators for sieving byproducts and, of course, threshing product.

Setaria tomentosa *rachillas*

Setaria tomentosa rachillas occur significantly in sieving product and less significantly in winnowing by shaking product (Table 4-8). Thus, the rachillas are cleaned initially in the sieving, but primarily in the subsequent winnowing by shaking process. Therefore, *Setaria tomentosa* rachillas can be indicators of these two processes' byproducts.

Setaria tomentosa *seeds*

Setaria tomentosa grains/seeds occur in significant frequencies in sieving product, and winnowing by shaking byproduct and product. As in the case of *Sorghum,* their presence in high frequencies in the sieving and winnowing by shaking products is expected, and the high numbers in winnowing by shaking byproduct and moderately numbers in sieving byproduct are unanticipated. Therefore, the *Setaria tomentosa* in the winnowing byproduct, and also in the sieving byproduct, were examined for distinguishing characteristics. As in the case of *Sorghum* grains, the majority of the *Setaria tomentosa* seeds from winnowing by shaking byproduct were either tail grains or were inflected by fungus or were rotten. Most of the grains in sieving byproduct had rotted badly and had swelled in size.

Weeds

The twenty different weed species of the *Setaria tomentosa* plot are present in moderate amounts and are categorized into 7 categories (see Table 4-8). The categories present were Big Headed Light (BHL), Big Headed Heavy (BHH), Big Free Light (BFL), Small Free Light (SFL), Small Headed Heavy (SHH), Small Free Heavy (SFH), and Small Headed Light (SHL). All categories are represented except the big free heavy weed category (BFH). Of all the categories, the small free light (SFL) and small free heavy (SFH) dominate the weed assemblage (86%), followed by big headed light (14%).

The other categories, big headed heavy (BHH), big free light (BFL), small headed light (SHL), and small headed heavy (SHH), occur in low frequencies (.3%). In general, the free weed seeds occur in significantly higher frequencies (86%) than the headed weeds (14%). This reflects the efficiency of the threshing process.

The patterning of the weed categories down the processing pathway is intricate in that the different categories occur in differing frequencies at different stage products and byproducts. For example, small headed light weeds (SHL), big free light weeds (BFL), and big headed heavy weeds (BHH) occur only in sieving byproduct. The small free heavy weeds (SFH) are present in all the stages byproducts and products, but occur in significantly higher frequencies only in winnowing byproduct. This reflects their removal through the winnowing process. Small free light weeds (SFL) remain into the winnowing product, and are removed only through later winnowings. Thus, the distribution of weed categories occurs in varying levels of significance at different processing stages, and therefore they can be used as indicators of these processing stages products and byproducts.

A Model for the Processing of Type B Crop with weeds (*Setaria tomentosa*)

The data regarding the patterns of component distribution are summarized in Table 4-9. Their distributions were collapsed into five ranges, which were, given qualitative terms: absent, rare, moderate, strong, and dominant. This table provides the skeleton for modeling crop processing of a Type B crop such as *Setaria tomentosa*, through threshing with sticks, sieving, and winnowing by shaking.

The four factors that are instrumental in assessing the probability of these components entering into archaeological contexts are then considered. They are fodder use (immediate or stored), storage of crop product (likely or definitely), prone to fire (catastrophic and intentional; degree of very likely to low likelihood), and lastly incorporation into dung as temper or fodder and subsequent use as fuel.

In the case of husked crop grains such as the *Setaria tomentosa,* storability is optimal until the pounding stage when the grains are dehusked. So, the storability index (NSLP, LSLP, or DSLP) for

Table 4-8. *Setaria tomentosa*. Crop Processing Stages Composition.

Standardized Counts for *Setaria tomentosa* (per 500 g for each stage)

	Sieving Product	**Sieving Byproduct**	**Winnowing by shaking Product**	**Winnowing by shaking Byproduct**
Straw (vol)	7428	82970	–	12500
Sorghum rachillae w/spikelets	–	1500	–	–
Sorghum inflorescence heads	29	1872	–	–
Sorghum w/husk (spikelets)	–	–	500	–
Sorghum grain	4142	137	5292	600
Setaria rachillas	9943	954	–	14450
Setaria seeds	3483914.2	9926	1809035	623300
Weeds Total	407337	23607	208247	91200
Total SFH Weeds	107052	17584	88763	79500
Total SHH Weeds	1486	46	530	500
Total SFL Weeds	206172	409	118008	11200
Total SHL Weeds	92571	5228	946	–
Total BFL Weeds	57	250	–	–
Total BHH Weeds	–	91	–	–

SFH=Small Free Heavy; SHH=Small Headed Heavy; SFL= Small Free Light; SHL= Small Headed Light
BFL= Big Free Light; BHH= Big Headed Heavy

Table 4-9. Tabulation of *Setaria tomentosa* distributions.

Stages	**Straw**	**Sorghum panicles**	**Sorghum grain w/o husk**	**Sorghum grain**	**Setaria rachillas w/spikelets**	**Setaria grain w/husk**	**Setaria rachillas**	**Weeds**	**BHL**	**BHH**	**BFL**	**SFL**	**SHH**	**SFH**	**SHL**
Sieving Byproduct	★★★	☆	–	☆	☆	★	☆	★★	★★	☆	☆	☆	–	★★★	☆
Sieving Product	☆	☆	–	☆	–	★★★	☆	★	★★	–	☆	★★	☆	★★	–
Winnowing by shaking Byproduct	☆	–	–	☆	–	★★★	☆	★	–	–	–	★	☆	★★★	–
Winnowing by shaking Product	–	–	☆	☆	–	★★★	–	★	☆	–	–	★★	☆	★★	–

BHL = big headed light; BHH = big headed heavy; BFL = big free light; SFL = small free light;SHH = small headed heavy; SFH = small free heavy; SHL = small headed light

– = Absent
☆ = Rare
★ = Moderate
★★ = Strong
★★★ = Dominant

millet crops such as *Setaria tomentosa* and *Panicum miliare* would also be a good indicator for when a crop could be best transported distances, i.e., to markets. The degree of exposure of a crop product or byproduct to fire, catastrophic and/or intentional, is probably the most important factor that determines the survival of the assemblage(s) into the archaeological record.

A model for the processing of *Setaria tomentosa* is presented in Figure 4-13. The processing pathway is simple with harvesting, raking, threshing, sieving, winnowing by shaking (1), and winnowing by shaking (2) being the main processing stages. There are no variations in the pathway; however, there are additional processing stages after winnowing by shaking (2) which are not presented in the model. These additional stages include a series of additional winnowing by shaking stages, which clean out the crop product more efficiently each time. The successive winnowing by shaking processes also separate the *Sorghum* and *Setaria tomentosa*. For the latter, winnowing by shaking is followed by pounding when the grains are dehusked. Often at this stage, the grains are roasted to facilitate the removal of the husks during the pounding process. Then, another winnowing by shaking removes the chaff-husk from the dehusked grain. Since the grain can be consumed in both grain and flour form, grinding occurs only when the consumer needs flour. In the case of *Sorghum,* the grain is ground to a flour after winnowing by shaking and before consumption.

From this processing model for a Type B plant (in this case *Setaria tomentosa*) it is evident that the smaller the crop grain seed being selected for in the crop processing stages, the higher the number of smaller sized weeds left behind for later crop processing stages. This pattern occurs because these smaller weeds are difficult to process out from crop seeds particularly when they have other properties in common. Big weed seeds are absent in the winnowing product, and small headed weed seeds are rare. In contrast, small free weed seeds, which are closest to *Setaria tomentosa* in morphology, are left behind with the crop grains. Thus, the remaining small free weeds are processed out in the last few series of winnowing I, and this is very important for archaeological modeling. Therefore, the weed seeds that most closely resemble the crop seed are the last to be processed out. Using this information about the Small Free Heavy weeds similarity to *Setaria tomentosa*, if there were big, free, heavy weeds such as *Cordiosperum* or *Lannea coromandelica* present in this sample, then I would predict they would remain into the winnowing product since they are similar in morphology to *Sorghum* grain.

Lastly, it is apparent that a processor does not have to be selecting for only one type of grain (i.e., only *Sorghum*). This study demonstrates that processing can occur reasonably efficiently when the processor is selecting for two different types of grains with different morphological characteristics in the same sample. The efficiency of the processing is definitely lower than if they were two separate samples, but it is done with acceptable efficiency. This observation has important archaeological implications. For example, if one recovers an assemblage with high concentrations of two different plant species, one cannot immediately conclude that one species is necessarily a weed. In the case of *Sorghum* and *Setaria tomentosa* (where *Setaria* is not considered a crop), *Sorghum* could be used as a staple crop and *Setaria tomentosa* could be used as starvation food. Since they grow together, they were harvested and processed together, and separated only in the last few processing stages. *Setaria tomentosa* is also used as fodder for cattle.

Implications for Archaeological Interpretation

This concluding section of the chapter discusses two methodological issues, summarizes the salient aspects of the models, and discusses the variations in processing when crops (both Type A and Type B) are cultivated for three different purposes; for only food, for food and fodder, or for only fodder.

Methodological Issues

There are two main methodological issues relevant to archaeological modeling that need to be clarified: (1) possible differences in processing pathways and their implications to the models presented; and (2) the effect of formation processes on the preservation of specific components, and their distribution in different sieve sizes.

Variations in Processing Pathways

Variations in crop processing occur within each processing stage of the different crops, but the sequence of stages in the pathway does not vary. Thus, threshing is always followed by sieving, which is followed by winnowing, and so on. There

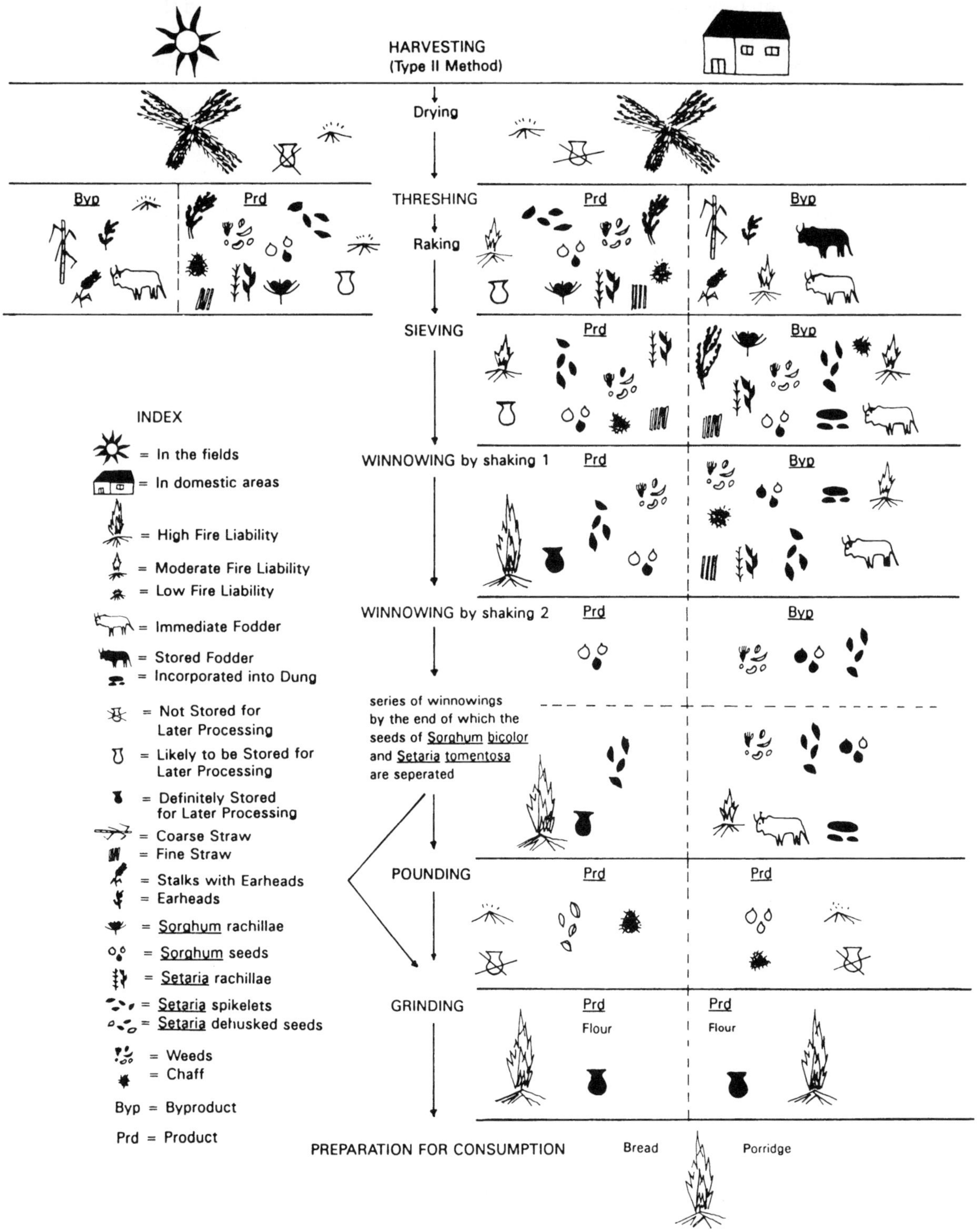

Fig. 4-13. Crop processing model for *Setaria tomentosa.*

is no variation in the course of the pathway itself, but there is a choice of what type of processing one does within one processing stage. For example, threshing can be done by cattle, sticks, or by feet. Winnowing can be done by wind or by shaking. A processor's choice of a certain type of threshing or winnowing is based on several factors including the type of crop being processed, amount of material being processed, availability of processing tools and man power, and wind velocity for winnowing by wind.

The ethnographic models developed for processing of various crops include different types of threshing and winnowing. However, if threshing and winnowing differed archaeologically from those presented in the model, one might suggest that this would affect the modeling. Careful consideration of the parameters of processing methods reveals that a distinct model need not be developed for each particular situation. The models developed are flexible to accommodate variations in the processing. In addition, such differences should have observable indicators in the various threshing and winnowing product and byproduct.

Threshing can be done through the use of cattle, sticks, and rubbing of the panicle inflorescence heads between the feet. These three threshing methods are different activities and this affects the material being processed (refer to 'Opportunistic' cultivations of ROC and SOC, which were threshed by two different methods). The most important difference is in the quantities of rachillas that are broken off during the different threshing processes. Threshing by sticks and cattle result in more breakage of the panicles and rachii and also a greater loss of crop grain in the byproduct, which then means higher frequencies of rachillas in the successive processing stages. Threshing through rubbing of panicles between the feet results in fewer rachillas in the product and also less loss of grain in the byproduct. As discussed for 'Opportunistic' cultivation, however, the selection of the threshing is based the on nature of the crop plant, particularly whether it is dry or green.

Winnowing can be done through the use of wind and through shaking. Processing through winnowing by shaking is a more directed intensive separation process as compared to winnowing by wind. It is important to note that small seeded crops are not winnowed by wind, since there would be a resulting loss of crop grain. Thus, it is mostly the larger and medium sized grain that is winnowed by wind. In India, the larger grained *Sorghum bicolor* and *Pennisetum typhoides* are winnowed by wind in addition to winnowing by shaking, while the smaller grained *Panicum miliare* and *Setaria tomentosa* are winnowed only by shaking. Therefore, when dealing with a small sized crop grain, it can be automatically deduced that it was *not* winnowed through wind.

Therefore, sorting out the effects of winnowing by wind and through shaking on the stage processing is only relevant to the crop plants with bigger sized grains such as *Sorghum bicolor* and *Pennisetum typhoides.* I suggest that in a majority of cases the first winnowing (usually directly following threshing in the fields) is always by wind, unless there are extenuating circumstances making it impossible. When threshing products are first winnowed by wind, the processing is more efficient than if they were only winnowed through shaking. Doing winnowing by wind first makes the successive winnowings by shaking more efficient, since it cleans most of the chaff, straw, and inflorescence heads. Such cleaning by shaking would be more time consuming. These components are easier to clean out through winnowing by wind because they are light and their separation is facilitated efficiently by the wind. If winnowing by shaking was the initial winnowing then the separation of these components becomes very labor-intensive since shaking (tossing and tapping) is not as effective in separating light materials.

When winnowing by wind and shaking are not done as successive processes, the compositions of the byproducts are distinctive and can to differentiated. The winnowing by wind byproduct has more prime grain (there is a greater loss of grain), and also more small weeds. Winnowing by shaking byproduct, however, has fewer prime grains, more tail grains, and fewer small weeds (because they are harder to process out through shaking). It is important to note that winnowing by wind will never replace any later winnowing by shaking processes. The only processing stage where there could be a choice of winnowing is the first winnowing. However, it is very rare that winnowing by wind in this initial stage is replaced through winnowing by shaking. Thus, new models need not be developed for variations in processing types but the extant model is flexible and can be altered for different situations.

Charring Experiment

The effect of formation processes on the preservation and distribution of components such as rachii and rachillas is an issue of concern. The distribution of the rachillas and rachii between sieves of different sizes (sieve A, sieve B, sieve C) is noted as an important distinguishing characteristic in certain models. However, questions could be raised regarding their relevance for the interpretation of archaeological assemblages. Formation processes (specifically carbonization through burning) might alter the frequencies of different sized components. For example, sieve A size components might be broken down into sieve B and C, and so on. Therefore, the use of sieve sizes as a parameter in the models might not always be practical given the nature of the particular archaeological assemblage.

To address this issue, an experiment was conducted to determine the survival of the rachillas and rachii upon burning/charring. Rachii and rachillas of *Panicum miliare, Setaria tomentosa, Sorghum bicolor,* and *Pennisetum typhoides* were charred in the laboratory ovens at high (450° C), medium (300° C), and low (150° C) temperatures. The rachii and rachillas of the three different sieve sizes were charred separately, and the different sized rachii and rachillas were charred in three distinct contexts (with sand in earthenware, without sand in earthenware, and without sand in aluminum foil). This latter portion of the experiment was done to facilitate variations in charring contexts. After charring was completed, the different sieve size components were examined for breakage and alteration. The results revealed that size transformations did not occur in significant amounts, regardless of the context in which the charring of the rachii and rachillas occurred. However, the main alteration occurred when charring at very high temperatures (over 450° C), which resulted in some sieve A components breaking down to sieve B sizes. The results of the burning experiment suggest that the integrity of the sieve sizes is a potentially viable archaeological criterion. Therefore, variations of components between sieve sizes can assist in the discrimination of products and byproducts of processing stages.

The question of the survival of the different types of *Panicum miliare* rachillas (without spikelets, with less than 50% spikelets, and with more than 50% spikelets) is similar to the discussion of the survival of different sized rachillas and rachii. The archaeological relevance of such a detailed classification of the rachillas could be questioned, and to address this question charring experiments were also done on the three different types of *Panicum miliare* rachillas. The charring experiment involved testing the survival of the rachillas at different temperatures (high, medium, low) and in different mediums (with sand in earthenware, without sand in earthenware, without sand in aluminum foil). After charring was completed, the three rachillas categories were examined for breakage and alteration.

The results revealed that transformations occurred at the high temperatures for the rachillas with spikelets. The rachillas with more than 50% spikelets were most affected since they lost enough spikelets in the burning experiment that they had to be reclassified into the less than 50% spikelets group. The rachillas with less than 50% spikelets were affected minimally, while the rachillas without spikelets were not affected at all. Certainly, two distinct groups of components, rachillas with spikelets and rachillas without spikelets, could still be distinguished. Thus, the burning experiment reveals that although the integrity of the rachillas with more than 50% spikelets tends to be affected at high temperatures, frequently variations between rachillas with and without spikelets can assist in the discrimination of the products and byproducts of the processing stages.

The survival of the different components of various sizes is also dependent on other processes unrelated to carbonization. The survival and preservation levels of organic material vary depending on the environmental context and the type of sediment in which they occur. Although these processes cannot be changed, the archaeologist and archaeobotanists can exercise control when excavating organic rich deposits. Careful and cautious excavations are critical for the recovery of fragile plant remains, and the excavator's knowledge of the physical appearance of archaeobotanical remains in sediments is crucial for successful recovery. In addition, processing of archaeobotanical remains should be cautious and conservative to reduce the fragmentation and post excavation alteration of the archaeobotanical materials. Without such care the alteration of archaeobotanical remains during excavation and processing will be of an equal magnitude to the modification done by formation processes.

Summary of Ethnographic Modeling

The ethnoarchaeological research conducted in India has established that there are distinct processing methods and pathways when a crop is processed for its grain. The selection of specific processing methods and pathways is dependent on whether the plant is a Type A crop or a Type B crop.

The difference between the two types of crops, Type A and Type B, lies in the morphology/structure of the plants themselves and this has a direct effect on the type of harvesting method used on the specific type of plant.

Type A crop plants such as *Pennisetum typhoides* and *Sorghum bicolor* have thick stems/stalks and only a limited number (1-3) of crop plant stalks to be cut at a time. Additionally, the inflorescence heads of these crop plants are large distinct panicles, which can be harvested individually. Type A crop plants, therefore, can be harvested in two distinct ways: by cutting off only the inflorescence heads and processing them for grain (leaving the stalks in the field), or by cutting 1-3 plants at a time at the base of the plants and then processing them for grain. In this latter method, weed plants are included into the processing pathway, though there could be some selection against them during the cutting itself given the small number of plants selected to cut at one time.

Type B crop plants, such as *Panicum miliare* and *Setaria tomentosa,* are harvested in a Type II method. The panicle inflorescences of these crops are not as large and distinct as those of Type A crops, and they cannot be singly cut from the plant efficiently. The stalks/stems of these plants are not thick, and a number of plants can be held and cut at the base when harvesting. This is the only method of harvest for Type B crops, and given the nature of the cutting and the crop plants, weeds are included into the harvest, and selection against them during the harvest is not possible. This expected inclusion of weeds during harvest is markedly different from the typical Type I method harvest of Type A crops where weeds are often excluded.

Using the crop processing studies of *Sorghum bicolor* and *Pennisetum typhoides* models were developed for Type A crop processing (see Figures 4-7 and 4-8). The key aspect of the Type A crop-processing model is that the method of harvest has a distinct effect on the successive processing stages in terms of the composition of different products and byproducts. The first method, involving only the cutting of the panicle heads, results in a 'clean' harvest where weeds, stalks, and straw are entirely excluded from the harvested material. The rest of the plant (the stalks) is harvested subsequently depending on availability of manpower, its need as fodder, and the schedule for sowing of the next crop.

In the second method of harvest, when a Type A crop is harvested by cutting at the base of the plant (similar to a Type II harvest), weeds, stalks and straw are included in the harvested materials, and thus need to be processed out in the various stages of the pathway. The harvesting cut may occur either at the very base of the plants (leaving a very small stub which could be removed through ploughing), or the cut could occur up about 1/4 the height of the plant, leaving a longer stubble in the field for cattle to graze. These two cutting variants differentially include low-lying weed species in the harvest, but since no other differences exist, they are classified together under one method.

Often when Type A crops are harvested in a Type II method, they are dried on the threshing floor and occasionally the stalks are separated by cutting them off and directly feeding them to the animals. In such situations, the bulk of the material to be processed is removed, leaving the grain bearing heads and most of the weeds on the threshing floor. This method of processing still creates a very distinctive assemblage in contrast to when a Type A crop is harvested in a Type I method.

I propose that for archaeological applications the most relevant harvest method for a Type A crop is the Type II harvest method, when crop plants are cut at the base, thus incorporating the weeds. The thick stems and large sized panicles of Type A crops are probably related to the 20th century genetic alterations to facilitate greater yields of both grain for human food and stalks/straw for fodder usage. These hybrid varieties need a Type I method of harvest, but it is plausible that the ancient varieties might have had thinner stalks and smaller panicles much like the Type B crops. In the absence of genetic information for the ancient strains, it is not possible to comment further on this hypothesis, but it is important to be aware of both possibilities, harvest of Type A crop in Type I method and Type II method, when conducting archaeological research.

Using similar ethnographic crop processing studies of *Panicum miliare* and *Setaria tomentosa*, two models were developed for Type B crop processing (see Figures 4-11 and 4-13). Since Type B crops are always harvested in the Type II method, a key aspect is that there are no variations in the processing of these crops due to harvesting. One parameter, however, that could have a significant effect on the patterning of the crop product and byproduct composition is whether weeds are present in the fields or not. The two models include processing a Type B crop associated with weeds and a Type B crop without associated weeds. In the absence of weeds, other plant parts (such as rachillas) can be used as discriminating variables to assist in identifying processing stages' products and byproducts.

In addition to the relationships between harvesting methods and type of crops, differences in successive processing were noted for the different crops, specifically with respect to the winnowing and pounding processing stages. Of particular importance is that the winnowing method selected is dependent on the size of the crop grain. Small-grained crops are not winnowed by wind, but rather they are winnowed by shaking and this reduces the loss of grain during processing. Crops of larger sized grains witness both winnowing by wind and by shaking. In this study, the Type A crops *Sorghum bicolor* and *Pennisetum typhoides* are both larger sized grain crops, while the Type B crops of *Panicum miliare* and *Setaria tomentosa* are the smaller sized grain crops. Although this classification is convenient for this study, not all Type A crops have larger sized grain and not all Type B crops have small sized grain.

Pounding is most often done to dehull glumed crop grains while grinding is done to convert grain into flour. In this study, pounding was an important stage for the Type B crops *Panicum miliare* and *Setaria tomentosa* processing, which are both glumed (like glume wheats). However, the Type A crops *Sorghum bicolor* and *Pennisetum typhoides* are free threshing (similar to naked wheats), and do not need dehulling before consumption. Therefore, pounding is not an important stage for these Type A crops. When pounding is a processing stage for *Sorghum bicolor* and *Pennisetum typhoides* it mainly functions to complete the removal of husk from grain that did not occur during threshing dehusking (i.e., removal of the seed grain from the spikelet).

There are several general implications of these models that need to be emphasized. Each of these inferences relates to the patterning and distribution of the different components up the processing pathway. They include:

1. No one variable can be used as a discriminating variable, but a combination of components is needed and is much more reliable.

2. Different processing activities create distinct assemblages. For example, winnowing by wind mostly cleans out lighter materials such as straw, chaff and lighter weeds, while winnowing by shaking is more directed and depends on what the processor wants to select out (specifically crop tail grain and heavier weeds). Sieving, on the other hand, is very specific in processing out rachillas, rachii, and large seeds (which are infected and bloated).

3. Larger sized components are cleaned out in the early stages while smaller sized components are cleaned out in the later stages.

4. The smaller the crop grain, the higher the frequencies of weeds in the later winnowing by shaking products. Additionally, the larger the prime grain, the fewer the number of weeds in the later winnowing products. It is a matter of efficiency in cleaning and selecting against items that are not the size of the prime grain. The greater the differences in size between the prime grain and the weeds, the more efficient is cleaning in the initial winnowings by shaking.

5. The main cleaning of the prime crop grain (versus tail grain) occurs in the later stages. This pattern is very significant for archaeological discovery and survival into the archaeological record, since the later stages significantly occur in home bases.

6. Weeds seeds are strong discriminating variables, but in their absence, rachillas can be used (as was done for the *Panicum miliare* processing model).

7. The probability of the different components entering the archaeological record is based on several factors. The important ones considered in this study were fodder usage, storage index, exposure to fire, and incorporation into dung.

8. A processor need not have to select only for one crop grain. As shown in the study of *Setaria tomentosa,* processing can accommodate selection for two crop grains within one sample (*Setaria tomentosa* and *Sorghum bicolor*).

9. Decision making plays an important role in the structure of the processing pathway. This is particularly noteworthy in terms of determining when a byproduct has too much crop grain and needs to be processed further along another parallel pathway, or when a byproduct becomes immediate fodder or stored fodder. The decision-making is based on experience of the processor, ethnic/sociocultural behaviors, and other limiting factors such as economic situations and work power.

In addition to developing models for the processing of the different types of millets (Type A and Type B), this research also proposed to model variations in crop processing when crops are cultivated for three purposes: for only food, for food and fodder or for only fodder. Significant differences in processing were observed for each of these three situations, and they varied depending on the crop type (Type A or Type B). These differences have measurable effects on the compositions of the products and byproducts of the different stages in the processing pathway (these differences were discussed earlier in this chapter). The differences in processing of a Type A crop when cultivated for the three purposes will be briefly summarized, and then the differences in processing of Type B crop when cultivated for the same three purposes are examined.

In the rare cases when a Type A crop is only cultivated for food, the harvest method involves only the cutting of the inflorescence heads. This results in a 'clean' harvest with no weeds, straw, or stalks (as discussed earlier). The focus of the cultivation and processing is only on the crop grain; therefore the harvest and processing only include the panicle heads. The resulting compositions of the processing stages products and byproducts are very distinctive, and primarily characterized by an absence of weeds, straw, and stalks. There is minimal concern for the stalks left in the field, and often they are left there for relatively long periods of time since they are not used as fodder. The fields are then cleaned before the next sowing. The effort involved in the harvest is moderate in comparison to when the crop is cultivated for the other two purposes.

When a Type A millet is cultivated both for food for humans and fodder for animals, then the harvest can be done in a method similar to Type II method, with the plants cut at the base. When a crop (Type A or B) is cultivated for both human food and animal fodder, the crop grain is used as the human food and the rest of the plant is used as fodder. The focus of the processing is still on retrieval of the crop grain for human consumption, but the rest of the plant (i.e., the stalks) is efficiently extracted for use as fodder utilizing the byproducts of the various stages. This is the most intensive processing in terms of effort invested when compared to the other two methods of processing (only for food or fodder). It is important to note that weeding is done in both cases when crops are cultivated for food and fodder, and for food only.

When a Type A crop is cultivated for only fodder, in most cases it is not allowed to seed; that is, it is harvested when green. Harvest is by a Type II method with the crop plants cut at the base incorporating all the weeds growing in the vicinity. Weeding is absent when a crop is cultivated for fodder. When a crop is cultivated only as a fodder, it is not processed further than harvesting, since the best cattle fodder is green fodder (only pregnant cows are given grain). Thus, the differences in processing of a Type A crop when it is cultivated for only food (in rare cases), for food and fodder (most common), and for fodder only, lie in the harvest methods and times. The different harvest methods have obvious, measurable effects on the composition of the products and byproducts of the various stages. The harvest time and whether a crop is processed or not (as is the case when a crop is cultivated only for fodder) have a direct effect on whether the cultivation is detectable archaeologically.

Differences in processing when Type B crop is cultivated for the three purposes were discerned, but the differences were not as distinct as those for Type A crops. The main difference lies in the harvest times, and as discussed earlier when a crop is cultivated for fodder only it is not processed after harvesting. When a Type B crop is cultivated only for food, or for food and fodder there is no significant difference in the processing except that in the former case byproducts are not fed as fodder to animals. There is no distinctive difference in the processing or harvest method between the two cases. It is, however, probable that the crop might be harvested slightly greener when the use is as fodder. This is important since it affects the compositions of the crop products and byproducts in the processing stages, particularly in terms of the frequencies of the rachillas. The greener the crop at harvest, the fewer rachillas in the crop products and byproducts (refer to the differences between ROC and SOC plots in the study of 'Opportunistic' cultivation of *Panicum miliare*). It could also be deduced that there would be more headed weeds in early harvests (when crop is still partially green) since the weeds are also immature and would not thresh free as well as when they were matured and dry. Thus, these two patterns could be used to determine whether a crop was cultivated and processed only for human food or for food and fodder.

When a Type B crop is cultivated only for fodder, then it is harvested green and not allowed to seed (as is the case for Type A crops). It is also not processed beyond harvesting. Thus, such cultivation is very unlikely to be detected archaeologically, and parallel lines of evidence are needed to determine such cultivation of plants for fodder.

Conclusions

This chapter has isolated the variables that distinguish the different crop processing stages' products and byproducts, and presented ethnographic models for archaeological interpretations. These models complement and extend the ethnographic models presented by Hillman (1981, 1984), Jones et. al. (1986), and Jones (1988, 1987) in that they focus on different crops and different processing methods. Hillman (1981, 1984) and Jones (1988, 1987), suggest that alternative processing methods are minimal, and therefore they lose the opportunity to discuss technological style in crop processing. For example, the necessary separation of grain from weed seeds and chaff is a processing stage with alternative solutions. The Near Eastern method predominantly involves sieving for this separation (Hillman 1984). As demonstrated in this study, however, a flat basket is tapped or shaken in South Asia, while in Australia a wooden dish is rocked (Cane 1989). This study has also demonstrated that the specific technique of winnowing (by shaking) varies significantly from the technique of winnowing by wind (which is the only winnowing method in the Near East). Furthermore, it has been demonstrated that the products and byproducts of winnowing by shaking are distinct from those of winnowing by wind.

Since these methods of separation produce statistically distinct product and by-product assemblages, it would be interesting to trace the development and adoption of different methodologies. Such research would be of particular interest in areas where both methods are known, yet one is specifically chosen, perhaps as an element of group identity. Jones (1984) has suggested that in order to identify activity areas, the functions of building etc., it is more important to know the processing stage rather than the processing method used. I argue that this approach has an important shortcoming relating to aspects of ethnic group identity and socio-cultural behavioral practices related to adoption of processing methods. There is obvious selection of processing methods and pathways based on several factors as discussed in this chapter, but ethnic and socio-cultural identity also play an important role. Therefore, whether it is a matter of specific crops needing different processing, or socio-cultural decision-making and determination, these differences could have wide ramifications in the study of ancient farming practices. This research has noted that different crops need different processing, but decision making on the specifics of the processing is identified as an important aspect of crop processing in South Asia.

Identification of reaping heights from charred remains has been discussed by van Zeist (1984) and Hillman (1981, 1984). They addressed reaping (which could be argued as being identical to Type II method of harvesting in this study) and uprooting (which was not observed in this study). However, millets were not considered in their research, nor was variation in the harvesting methods addressed. This study is an important new contribution due to its focus on millets and differentiation of crops on the basis of the harvest style. Moreover, an important implication of this study is the observation that distinctions between harvesting styles (such as Type I method and Type II method) have a direct effect on the compositions of the successive products and byproducts down the processing pathway.

5. Going Beyond Carbonized Seed Lists: Paleoethnobotanical Research

This chapter presents the paleoethnobotanical research at Oriyo Timbo and Babar Kot in Gujarat, India. The objectives of the archaeobotanical research at these two sites include the description and interpretation of the archaeobotanical materials recovered, and presentation of a working model for the economic role of millets in the subsistence system at each site and for the region as a whole. Gaining an understanding of the nature of the subsistence economy, and elucidating spatial and temporal variation in the plant materials at each site are important themes of this chapter.

First an archaeological background of the two sites is presented followed by a discussion of the archaeobotanical procedures employed in the study, including sampling, recovery, analysis, and identification. Finally, the results of the archaeobotanical analysis of materials from Oriyo Timbo and then Babar Kot will be presented.

Archaeological Background of Oriyo Timbo and Babar Kot

The archaeobotanical field investigations at Oriyo Timbo and Babar Kot were conducted by the author as a part of larger excavation projects. Both the sites are located in the modern state of Gujarat in northwest India (Figure 5-1), and were excavated through collaboration between the Gujarat State Department of Archaeology and the University of Pennsylvania, under the direction of Dr. Gregory L. Possehl. Oriyo Timbo was first excavated in 1981-82 by the joint team (Rissman and Chitalwala 1990), and a subsequent season of excavations took place in 1989-90. The archaeobotanical research discussed in this book was conducted during this second season at Oriyo Timbo (1989-90). The site of Babar Kot was excavated in 1990-91 by a similar joint team collaboration between the Gujarat State Department of Archaeology and the University of Pennsylvania, under the direction of Dr. Gregory L. Possehl.

Oriyo Timbo

Oriyo Timbo has Lustrous Red Ware ceramics and radiocarbon dates indicating Late Harappan occupation around 1900-1800 B.C (Chitalwala and Rissman 1990). Below the Late Harappan occupation is an earlier microlithic occupation, which was not well defined in the first season of excavations and not explored at all in the second season. Oriyo Timbo is located inland on a cultivated plain between two rivers. The Ghelo River is situated 8 km north of the site and the Khalubhar River lies 4 km south (Possehl 1980:38). The site itself (a low mound, 150 by 160 m) sits on the banks of a seasonal tributary of the Khalubhar River.

The location of Oriyo Timbo away from a major river, and only along a seasonal tributary is of importance in reconstructing its role in the regional settlement pattern. There are numerous Harappan sites along the two nearest rivers (Possehl 1980), and from a sample of 36 Harappan sites only Oriyo Timbo is not located within 200 m of the Ghelo or Khalubhar Rivers. Possehl (1980) has suggested that Harappan site locations were typically near the river bank and were designed to take advantage of a strip of land kept free of dense vegetation by periodic flooding, thus eliminating the need to clear land for cultivation. Thus, the riverside location of Harappan sites was designed to facilitate cultivation. The location of Oriyo Timbo is intriguing since it obviously does not optimize this situation, therefore suggesting that flood plain agriculture might not have been a critical subsistence component; however there could very well be other factors that also determined the location of the settlement.

During the first season at the site, three excavation operations were conducted: the eastern, western, and central operations (Rissman and Chitalwala 1990: 5-11). The western operation was the most successful, and the excavation units of the second season focused around this location. As a result of the initial season of fieldwork, Rissman and Chitalwala (1990) suggested that Oriyo Timbo was not a self-contained permanent settlement. Instead it was occupied only on a seasonal basis, with the inhabitants living elsewhere at other times of the year. They concluded that Oriyo was an ephemeral site based on the presence of hearths, hoof prints, a possible windbreak tent, no permanent architectural features, and a very high density of herd animal (cattle) bones. Rissman (1985) investigated these indications of a non-agricultural, non-sedentary Late Harappan site through analysis of annual rings in cattle teeth to determine the times of years when they were slaughtered. The fauna from Oriyo Timbo exhibited a predominant season of death between March and July, suggesting that the livestock, and therefore perhaps humans did not occupy the site year round. However, the presence of millets (*Setaria* spp., *Eleusine* spp., *Panicum* spp.) at the site complicates this interpretation (Rissman & Chitalwala 1990; Wagner 1982). A key unresolved research question was whether the millets recovered at the site were cultivated there or brought in from elsewhere.

The second season of field research entailed the excavation of six 5 by 5 m grid units immediately south of the western operation area of 1981-82 excavations (Rissman and Chitalwala 1990:Figure 1). Approximately 3% of the site was excavated during the two seasons. Only two of the six trenches were excavated to sterile sediment. The others were dug down to expose a series of occupational surfaces with associated features (Possehl 1990). All soil (except the disturbed plough zone which was 20-30 cm deep) was screened for artifacts, and in addition large volumes of unscreened sediment were floated from all excavated areas for macrobotanical remains. Soil sediments from features were often floated in their entirety. The material culture from the site included a variety of ceramic types, majority of which were Lustrous Red Ware.

In the course of the 1989-90 excavations twelve stratigraphic layers and sub-layers, which comprised no more than 3 major distinguishable occupations, were identified. The depth of the cultural deposits ranged from 120 to 130 cm below the surface. The top 20 to 40 cm of the cultural material below surface were disturbed to varying degrees through agricultural activities. The upper layers of the site are associated with the topsoil and the plow zone, and are extremely disturbed. These strata (1 and 2) are distinct from stratum 3, which

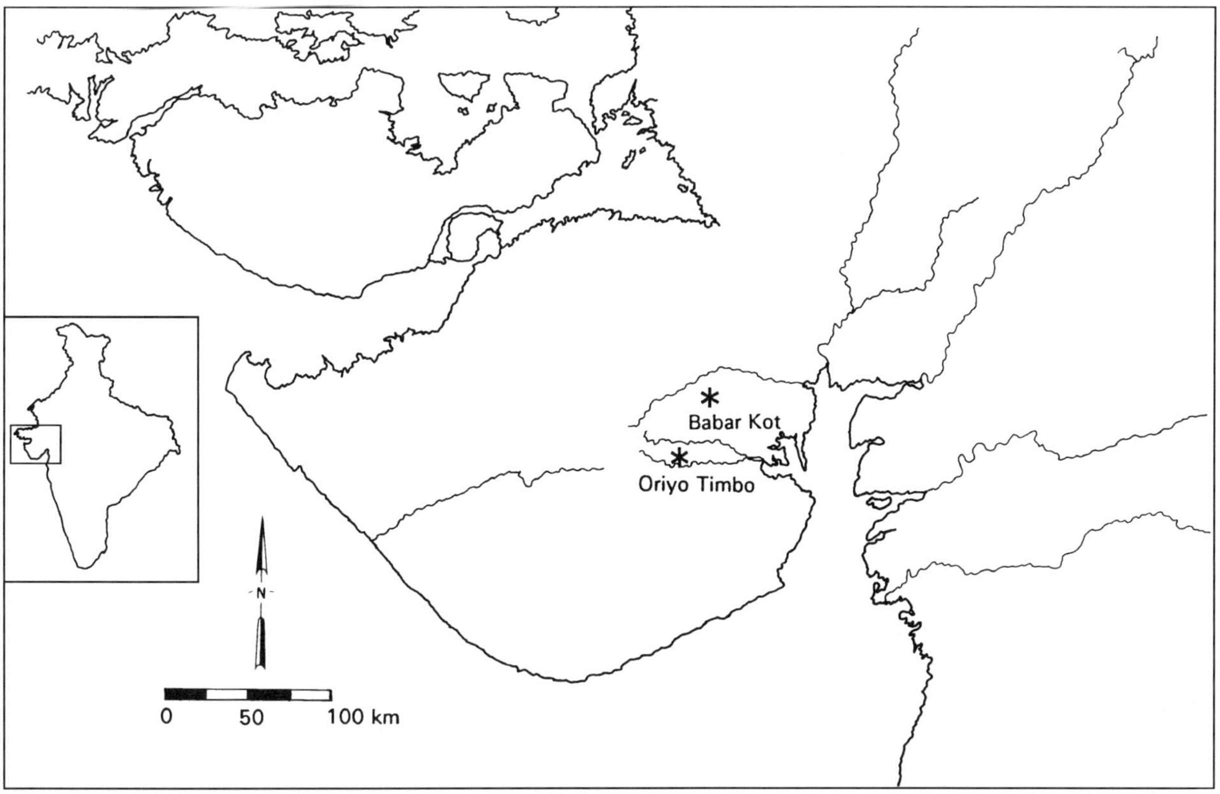

Fig. 5-1. Map showing the sites of Oriyo Timbo and Babar Kot.

is a series of very compacted soil layers and artifact concentrations. Stratum 3 is considered to represent the latest Late Harappan occupation surface at the site. Stratum 4, a softer compacted fill, is distinct from the compacted series of layers in stratum 5 (5, 5a, 5b). Associated with these layers are high densities of ceramics and features associated with food preparation (i.e., hearths, a tandoor oven, and ash pits). Stratum 6, a soft fill, separates the next series of compacted occupation surfaces (6a, 6b, 6c), which have occupation features, ceramics, and various artifact concentrations most closely, related to Lustrous Red Ware cultural sequence. These Late Harappan surfaces extend down directly onto the sterile soil of stratum 7. Stratigraphically, strata 6a, 6b, and 6c are associated with Occupation I, strata 5, 5a, and 5b are associated with Occupation II, and Occupation III is represented by stratum 3 (Possehl 1990). The disturbed upper strata and the unexcavated microlithic occupation are not considered further in this study. Based on the second season of excavations, the excavators propose that there are three distinct Lustrous Red Ware occupations, each associated with a series of compacted surfaces with features and concentrations of artifacts resting upon them. Softer, less compacted soil of strata 4 and 6 represents the general deposits between the three occupation layers.

The significant features at Oriyo Timbo included hearths, pits, and/or trash pits. Additionally, a tandoor oven was found in Occupation II, its inner walls charred and containing an ashy, charred matrix. The contents of this feature were floated to retrieve archaeobotanical materials. An interesting semi-spherical feature with sides sloping downward toward the center was observed in Occupation III. The central portion contained a compacted floor surface with a diameter of 2 to 3 m. Although a final interpretation has not been offered by the excavators, it is likely that it could be a living surface associated with a temporary structure, and the raised walls represent the perimeters of the dwelling.

My archaeobotanical research conducted during the second season of fieldwork was designed to elucidate the nature of the agricultural economy at Oriyo Timbo. Specifically, the research was designed to determine whether crops were cultivated at the site by the occupants or brought to the site from elsewhere, through the identification of lower and/or higher crop processing stages. In other words, an objective of the study was to determine whether analysis of the 1989-90 macrobotanical assemblage from Oriyo Timbo supports or contradicts earlier interpretations of Rissman (1985) and Rissman and Chitalwala (1990).

Babar Kot

Babar Kot is a very late Mature Harappan site set on the banks of a small seasonal river in the semi arid portion of Saurashtra, and a variety of crops are currently cultivated in the immediate environs (Figure 5-1). The site is a pronounced mound (190 by 140 m), about 2.7 hectares in area and 2.5 m high (Possehl and Raval, 1991). There were two excavation areas at the site: the main excavation area on the sloping southwestern quadrant of the site; and a second excavation area on the mound 50 m north of the main excavation area. This latter mound excavation consisted of a 5 by 10 m trench (Possehl and Raval 1991). The main excavation area is referred to as the slope area, while the 5 by 10 m trench on the mound is referred to as the mound area. A total of 28 excavation trenches, each 5 by 5 m in size, was opened in the two areas. The main excavation area exposed and explored a fortification wall and also entailed horizontal exposure of a habitation area with wall foundations. These wall foundations were of two types, stone and very clean, well-prepared black clay ("kali mate"). The full nature of the fortification is not understood at this point, but evidence suggests that it dates to the initial occupation of Babar Kot (Possehl and Raval 1991). Approximately 3% of the site was excavated in 1990-91.

The basic stratigraphy was similar in both areas. The top stratum produced early historic Red Polished Ware. These upper deposits (divided into strata 1 and 2) rest directly upon the Harappan occupation (Possehl and Raval 1991). The stratigraphic break was very well defined by the color and texture of the soil, thus the two principal occupations (Early Medieval and Harappan) are clearly demarcated. The excavators propose that the early historic period occupants certainly leveled the top portion of the Harappan site carefully before building their settlement. The trench on top of the mound, which has the greatest depth of cultural deposits as well as the least disturbance, reveals sterile soil at a depth of 3.25 m below the present ground surface. The occupation deposits consist of 11 different stratigraphic layers, which have been divided into three building phases (Possehl and Raval 1991). The building phases are numbered with Roman numerals from bottom to top with the bottom being the first phase of occupation. The order of the stratigraphic layers is reversed, with the numbering top to bottom and the top being the most recent stratum. Therefore, the early historic occupation, phase III is associated with strata 1 and 2. Phases II and I are interpreted as Harappan based on the presence of Harappan ceramics. Phase II is comprised of strata 3, 4, 4a, 5, and 5b, while Phase I (the earliest building phase) includes strata 7 and 8. Thus, Occupation I is the earliest and is associated with the strata 9 and 10. It has no buildings and is considered to be the initial occupation. Occupation II is associated with strata 6, 7 and 8, and corresponds to the building phase I. Occupation III consists of strata 3, 4, 4a, 4b, 5, and 5a, and is associated with building phase II. The last occupation (building phase III) was not studied for this project, since it is historic in age. Occupation II (strata 5, 5A and 5B) consists of a compacted series of occupation floors, which have food preparation, features such as hearths, ovens, and ash pits (Possehl 1990). Similarly, Occupation I (strata 6A, 6B and 6C) comprises occupation surfaces with high densities of ceramics and other domestic features. Strata 6 and 4 are softer fills separating the different occupation events. Occupation III (the middle building phase, phase II) is the most extensive and on the mound is represented by a well-preserved stone wall dividing the excavated area into two rooms (Possehl and Raval 1991). The cultural material associated with this architectural phase is "Rojdi C" in style. The corner of one room in this phase has remnants of a hearth or cooking area. A feature related to storage was recovered from the corner of another room. Soil samples were taken from these contexts to retrieve organic remains. Occupation II (the earliest architectural phase) has fewer features and it is difficult to speculate much about the buildings associated with this phase (Possehl and Raval 1991). The associated cultural remains differ slightly from the Occupation III materials.

Radiocarbon dating of Occupation I and II at the site yielded several dates which place site occupation between 2200-2050 B.C. cal (Herman 1997a; Possehl 1992). There appears to be some ambiguity in assigning Babar Kot to Mature or Late Harappan, since Possehl (1992:144) equates it to Rojdi C which is post-urban, while Herman (1997a:Table 8) classifies it as a Mature Harappan site. In this book, Babar Kot is characterized as a very late Mature Harappan site occupied at the time of transition into Late Harappan (which dates to 2000-2050 B.C. according to Herman [1997a:101]).

Summary

The Harappan sites of Oriyo Timbo and Babar Kot, present two very different archaeological contexts to study the role of millet cultivation during the late third and early second millennium B.C. in Gujarat. Oriyo Timbo had been previously interpreted as an ephemeral camp where the focus of the subsistence economy was animal husbandry (Rissman and Chitalwala 1990), while Babar Kot has substantial architecture and extended period of occupation. Among the questions to be addressed by the analysis of the archaeobotanical materials and the application of the new models of ethnographic crop cultivation were: did the inhabitants at both sites practice agriculture; were crops processed at each site; were millets cultivated for different purposes at the two sites; were there different emphases on different millets; and can any pattern in crop cultivation and millet emphasis be derived from the late Mature Harappan to the Late Harappan based on these two sites?

Archaeobotanical Methods and Sampling Procedures

The archaeobotanical methodology employed in this project emphasizes the importance of intensive sampling and the use of an efficient plant recovery system. Sampling and macro-botanical recovery methods were held constant in the investigations at Oriyo Timbo and Babar Kot. This was done to facilitate comparability between the two data sets for patterns of distribution, occurrence, and ubiquity of the archaeological plant remains. This is necessary to address issues related to variation and similarity in millet cultivation at the two sites.

Sampling

At Oriyo Timbo, 3,139 liters of archaeological sediment from 247 samples were processed, with a minimum volume of 10 liters per sample unless there was a defined feature. Similarly a total of over 200 soil samples representing over 3,000 liters were floated for archaeobotanical remains at Babar Kot. The sample contexts from the two sites include hearths, ovens, compact living floors, dumps, pits, and fill areas. Sediment samples from deposits that did not have any features were also taken for flotation. Essentially, floatation samples were collected as often as possible. As noted, such sampling methods are necessary for the successful recovery and identification of macro-assemblages and localized associations related to different economic activities such as plant processing, consumption, and disposal. Given this rigorous sampling it is asserted that the sediment samples floated for archaeobotanical remains are very representative of the excavated areas of the two sites.

Recovery

The water flotation system used for plant recovery comprised of an oil tank modified to an S.M.A.P-type machine (called Piyush One) used previously at the sites of Rojdi and Oriyo Timbo (Wagner 1985; Weber 1989). Recovery tests run with charred poppy seeds indicate that 86-97% of floating seeds were recovered. Special emphasis was made on having standardized control on the volume of sediment processed. A standard of 10 liters was taken for each float sample, unless it was from a feature, feature fill or a floor. As with most if not all plant recovery systems, the major concern during flotation was to minimize loss, contamination, and damage to the plant remains. The Oriyo Timbo (1989-90) and Babar Kot (1990-91) recovery methods minimized these concerns by thorough cleaning of the flotation tank, heavy fraction screen, and the light fraction box. The use of a continuous water supply automatically controlled for much of this.

Identification and Analysis

A comparative collection compiled in India by the author, Dr. Vishnu-Mittre, and Dr. Steve Weber was used for seed identification. Charcoal and carbonized seeds were separated from the light fraction samples by scanning under a light binocular dissecting microscope at 10X, 20X, and 40X. Both carbonized and uncarbonized seeds were recovered, however in this study a conservative stance is adopted and only the carbonized seeds are in the analysis. Since organic preservation is poor at the sites, the uncarbonized seeds are interpreted as later intrusives and not of Harappan context, and their exclusion increases the reliability of the assemblages. The only way to ascertain the age of these uncarbonized seeds would be through accelerator dating. The condition of the carbonized seeds varies from very good preservation to badly charred distortion. Drawings were made of the primary seed types to facilitate accurate identification. Select samples were analyzed in greater detail from both sites. The selection of samples for detailed analysis was based on several criteria, including strong Harappan context, minimal disturbance associated with the samples, and preference for samples from household activity contexts.

In order to more accurately examine patterns in the data, and to avoid any biases due to differences in volumes of sediment floated (Popper 1988:60), certain quantification methods are needed. For the study of archaeobotanical remains of Oriyo Timbo and Babar Kot, densities and percentages were chosen Densities are generally used as a measure for understanding depositional and preservational variability, and it is one of the most basic ratios used by paleoethnobotanists (Miller 1988:2). The basic assumption is that if everything is equal, the larger the volume of sediment floated the more plant remains will be extracted. In other words, the higher the density, the more intense the activities involving the particular plant and fire (Pearsall 1982:129). Therefore, with a good control and understanding of the context, the use of specific plants by site inhabitants can be reconstructed. The second quantification method used in this book is that of percentage which refers to the relative abundance of a specific taxon in an assemblage. Paleoethnobotanists use it as an indicator of floral importance (Miksicek 1983), or to compare different preservation contexts, or to detect replacement of one category by another (Miller 1988), or as a magnitude of past accidents (Minnis 1978). In this book, both densities and percentages are used for robust paleoethnobotanical reconstruction. Carbonized seed densities at Oriyo Timbo are reported as seeds per 1,000 liters of sediment, and for Babar Kot as seeds per 100 liters of sediment. The standardization for the Oriyo Timbo data was for a higher volume (1,000 liters) to allow for identification of subtle patterns since seed recovery at the site was lower than at Babar Kot.

Oriyo Timbo Archaeobotanical Results

The goal of the archaeobotanical research at the site was to elucidate the role of millets in the subsistence system, identify crop cultivation, and to present a working model for the economic role of millets in the subsistence system at Oriyo Timbo. The majority of the archaeobotanical materials recovered was scattered in the archaeological matrix of the site, although some samples came from quite distinguishable contexts such as ashy pits and features. The plant remains occur in varying frequencies in all the excavated trenches, but significantly higher concentrations were present in 3 of the 6 trenches. Stratigraphically there is patterned variation in the frequencies of carbonized seeds recovered. Occupation II produced higher frequencies of seeds as compared to Occupations III and I.

There is strong evidence for the use of crops including summer crop plants such as *Eleusine coracana*, *Panicum miliare*, *Setaria* spp. and legumes (*Phaseolus mungo* and *Brassica* sp.). Weeds including *Mollugo* sp., *Urochloa* sp., *Trianthema* sp., *Carex* sp., and *Euphorbia* sp. are also present in relatively high proportions. Whether the crops were cultivated at the site by the occupants or brought onto the site from elsewhere is an intriguing question, and the archaeobotanical analysis investigates this issue. Table 5-1 presents the spatial data from Oriyo Timbo. A total of 413 carbonized seeds have been identified, of which 41% are 'millets' (summer crops) and 51% are weeds. Among the millets, *Setaria* spp. comprises 16% of the total seed assemblage, followed by *Eleusine* sp. (15%), and *Panicum* spp. (8%). Legumes constitute 6% of the total assemblage. Among the weeds, *Mollugo* sp. (32%) comprises the highest percentage. No rachii or other plant parts were recovered. As compared to the archaeobotanical finds of the first season of excavations at Oriyo Timbo (Wagner 1982), the second season had higher overall recovery rates. The higher recovery rates are due to the larger quantity of sediment processed through flotation, and the use of a more efficient plant recovery system.

Millets

A total of 186 carbonized seeds belonging to three genera, *Eleucine* sp. (n=72), *Panicum* sp. (n=35) and *Setaria* spp. (74), commonly referred to as millets was recovered from Oriyo Timbo (Table 5-1). The *Eleusine* sp. caryopses were recovered from 4 trenches, with Trenches S2 and S6 having significantly higher densities. The 35 carbonized grains of *Panicum* sp. account for 8% of the total carbonized seed assemblage at the site. The 74 carbonized seeds identified as *Setaria* spp. were recovered from 4 of the 5 trenches, and are represented by three species: *Setaria italica*, *Setaria glauca,* and *Setaria tomentosa*. *Setaria italica* occurs in the highest frequencies, followed by *Setaria tomentosa*, and *Setaria glauca* occurs in lower frequencies. *Setaria italica* was most likely used as food most prominently among all the *Setaria* spp. represented at the site. Please refer to Reddy (1994) for further details.

Ethnographically, *Eleusine coracana*, *Panicum miliare,* and *Setaria italica* seeds are used primarily as human food. They are consumed either cooked like rice, or made into a flour for bread and porridge. The vegetative parts of the plant are valuable as green and dried fodder for animals. The storage of *Eleusine coracana* is extensive, and in majority of the cases, fresh grain is seldom used as food, but has to be stored for some months before it is utilized. *Panicum miliare* straw is also used as packing material due to its softness. The crop grains store well for long periods of time, both in a husked and dehusked state, and ethnographically it was noted that the storage period is relatively long for *Panicum miliare* (often up to a couple of years).

Legumes

A variety of legumes have been recovered from all 5 trenches, comprising 29 carbonized seeds (Table 5-1). The legumes occur in significantly higher densities in Occupation II of Trenches S2, S5, and S6 (Table 5-1). The presence of legumes, or any other seeds, in the archaeologically sterile stratum 7, is probably due to bioturbation and other related pedological activities. The legumes include two identified crops, those of *Phaseolus mungo* and *Brassica* sp. Six seeds were identified as being those most closely related to legumes, but further identification was not possible due to the fragmentary nature of the seeds. The diagnostic parts of these seeds, which facilitate identification, are missing.

Phaseolus mungo is also known as *Vigna angularis* or black gram. This legume is used as human food and the vegetative parts of the plant are popular as fodder. This legume is an important dry land crop in India, and is usually cultivated in low rainfall conditions (Aiyer 1982). It grows best in clayey and black cotton soils, and its presence indicates summer cultivation. The other legume crop represented is *Brassica* sp. Only one seed of this plant was recovered from Trench S6 Occupation I. *Brassica* sp. is a crop plant in India, and is often grown in widely different conditions of climate.

Non-Crop Plants

A total of 190 seeds were recovered which have been grouped as 'weeds' (Table 5-1). They are grouped together because they do not represent any major crops (as crops are defined today), although it is quite likely that some of them might have been specifically collected for food, for example *Amaranthus* sp., *Chenopodium* sp., *Dactyloctenium* sp., and *Solanum* sp., *Mollugo* sp. weed seeds occur in the highest frequencies, and they constitute the majority of the weed assemblage (Table 5-1). The weeds at Oriyo Timbo include 13 different genera of plants and represent a variant group, which could later be used for further interpretation of the archaeobotanical assemblage of Oriyo Timbo.

Table 5-1. Oriyo Timbo. Carbonized Seed Densities by Trench.

	S2	S5	S6	S8	S9	Total
Sediment Floated (liters)	320	200	404	175	576	1675
Carbonized Seeds (Absolute Counts)	106	15	139	80	73	413
Seed Densities (N) per 1000 liters	**S2**	**S5**	**S6**	**S8**	**S9**	**Total***
Carbonized Seeds	331	75	344	457	127	247
Eleusine sp.	81	-	79	17	19	43
Panicum miliare	56	15	30	6	2	21
Setaria spp. (total)	109	-	72	29	9	44
Setaria italica	63	-	42	11	5	25
Setaria glauca	-	-	7	-	-	2
Setaria tomentosa	47	-	22	17	4	17
Unidentifiable millet seed fragments	-	-	7	11	-	3
Total Carbonized millet seeds	246	15	188	63	30	111
Phaseolus mungo	28	25	12	-	5	13
Brassica sp.	-	-	2	-	-	0.6
Unidentifiable legume seed fragments	-	-	10	2	2	4
Total legumes	28	25	24	2	7	17
Amaranthus sp. (SFH)	3	5	3	-	-	2
Carex sp. (SFH)	3	-	27	-	-	7
Chenopodium sp. (SFH)	-	-	10	-	-	2
Cyperus sp. (SFL)	-	-	-	6	2	1
Dactyloctenium sp. (SFL)	-	-	2	-	-	0.6
Digera sp. (SFH)	3	-	-	6	4	2
Digitaria sp. (SFH)	6	-	10	-	-	4
Euphorbia sp. (SFH)	-	20	5	-	-	4
Mollugo sp. (SFH)	13	5	32	354	21	55
Phyllanthus sp. (SHH)	-	-	-	-	2	0.6
Solanum sp. (SFH)	3	5	5	6	-	3
Trianthema sp. (SFH)	-	-	12	11	7	7
Urochloa sp. (SFL)	-	-	25	6	56	26
Total Weeds	31	35	131	389	92	114

SFH = Small Free Heavy; SFL= Small Free Light; SHH = Small Headed Heavy

*per 1000 liters

These 13 different genera can be classified into the different weed categories developed in chapter 5. The resulting categorization reveals the presence of three weed categories, and the weed assemblage is dominated by small free-headed weeds (SFH) with about 69% of the weeds. The small free light (SFL) weed category constitutes 23%, while the small-headed heavy category (SHH) has a lower frequency of 8%. It is important to note that no big sized seeds are present, nor is the small headed light (SHL) category of weed seeds present. The ethnographic modeling has demonstrated that specific categories of weeds can be used as indicators of different stages of crop processing, and the categories present in the Oriyo Timbo weed assemblage (SFH, SFL, and SHH) will be used to shed light on what the archaeobotanical assemblage at the site is most indicative of: earlier processing stages or later processing stages. This information can be used to determine whether the crops were cultivated at the site or brought in through exchange from elsewhere.

Spatial Trends

Comparison of the carbonized seed densities between the five trenches reveals a distinct spatial patterning of the macrobotanical remains on Oriyo Timbo (Table 5-1). Trench S8 has the highest density followed by S6 and S2 with moderate densities, S9 has moderate to low density, while S5 has the lowest density of carbonized seeds. Trench S8 has the maximum number of features and ash pits, and it appears that there is a direct association between the context of the sample and the density of seeds recovered.

Millets (summer crops) occur in highest densities in Trench S2 and S6, while S8, S5, and S9 have significantly lower densities. The distribution of the genera is relatively consistent with the different millets occurring in similar frequencies in all the trenches (i.e., highest frequency in Trench S2, and lowest in S9). It is interesting to note *Eleusine* sp. and *Setaria* spp. are absent in Trench S5. *Panicum miliare* seeds were recovered from all the five trenches, but with Trenches S2 and S6 having higher densities (Table 5-1). Trench S2 also had the highest densities of *Setaria* spp. seeds while the other trenches have moderate to low densities of *Setaria* spp. (except Trench S5). Of the legumes, the 22 seeds fragments of *Phaseolus mungo* were recovered from 4 trenches, and they occur in higher densities in Trench S2 and S5. Weeds dominate the archaeobotanical assemblage of Trench S8, specifically the weed seeds of *Mollugo* sp. Weeds also occur in higher densities in comparison to crop plants particularly in Trenches S8 and S9.

Stratigraphic Patterns

Occupation II has the highest density of carbonized seeds (Table 5-2). Within Occupation II, stratum 5A has the highest density. Occupation I has lower densities of carbonized seeds but they are significantly higher than the densities of carbonized seeds in the fills between occupations. These results correlate well with the archaeological stratigraphy related to food preparation features.

Occupation II, specifically stratum 5A, has the highest crop seed density (millets and legumes) but low densities of weeds (Figure 5-2). *Eleusine* sp., *Panicum miliare*, and *Setaria* spp. occur in the highest frequency in stratum 5A of Occupation II. Even though *Eleusine* occurs in highest densities in Occupation I overall, stratum 5A of Occupation II has the highest density for this crop. Thus, its usage was higher in Occupation I overall as compared to Occupation II, even though one stratum of Occupation II has the highest densities of *Eleusine*. The high densities of *Eleusine* in Occupation II stratum 5A can be correlated to the fact that the food preparation features occur in these strata. *Eleusine* does not exhibit differential correlation to context, since it is almost equally distributed between samples from feature contexts and non-feature context. The distribution of *Panicum miliare* within Occupation II is comparable to that of *Eleusine coracana*, however its distribution in Occupation I strata is distinct from that of *Eleusine* in that it occurs in significantly lower densities. Within Occupation I, *Panicum miliare* is absent in strata 6A and 6C, and occurs in low densities in stratum 6B. Contextually, *Panicum miliare* seeds occur in higher densities in contexts not related to features (Table 5-2). In other words, their density is higher in general deposits. The distribution of *Panicum miliare* suggests that they occur in significantly higher densities in Occupation II as compared to Occupation I. Since *Panicum miliare* occurs in the strata associated with food preparation more significantly than those associated with occupation surface, this could be attributed to its status and role as a crop primarily related to human consumption. Stratigraphically, *Setaria* spp. occurs in the highest densities in Occupation II (Table 5-2). Within this occupation, stratum 5A has the higher density, while strata 5 and 5B have lower densities. Occupation I has relatively lower densities for *Setaria* spp. This distribution is similar to that of *Panicum miliare* in that the usage of these two millet crops is higher in Occupation II as compared to Occupation I, and this is in contrast to the usage of *Eleusine* sp. crop seeds which is higher in Occupation I. *Setaria* spp. occur in higher densities in strata (particularly 5A) which have features and contexts related to food preparation and disposal. Even though the seeds do not show higher densities or specific association with samples from features (Table 5-2), it is argued here that their significantly higher densities in the strata associated with food preparation suggest a role as food for human consumption.

Legumes, specifically *Phaseolus mungo,* also occur in the highest frequencies in stratum 5A of Occupation II. All the crop seeds (including millets and legumes) occur in significantly higher frequencies in stratum 5A of Occupation II. Again, as in the case of the millet seeds, the seed fragments of *Phaseolus mungo* do not occur in higher frequen-

Table 5-2. Oriyo Timbo. Carbonized Seed Densities by Occupation and Context.

(N per 1000 liters)	**Fill**	**Occupation II**			**Fill**	**Occupation I**			**Fill**	**Non-Features**	**Features**
Stratum	**4**	**5**	**5A**	**5B**	**6**	**6A**	**6B**	**6C**	**7**		
Carbonized Seeds (per 1000 L)	116	220	719	30	297	295	426	–	62	259	210
Eleusine sp.	22	23	193	–	43	159	125	–	–	48	39
Panicum miliare	9	10	140	–	26	–	42	–	–	26	13
Setaria spp. (total)	24	27	263	15	35	45	51	–	–	64	13
Setaria italica	9	16	158	15	26	45	10	–	–	37	6
Setaria glauca	–	1	–	–	9	–	10	–	–	2	2
Setaria tomentosa	15	10	105	–	–	–	31	–	–	25	5
Unident. millet seed fragments	3	2	–	–	9	–	21	–	–	2	6
Total Carbonized millets seeds	58	62	596	15	113	204	239	–	–	140	71
Phaseolus mungo	6	4	70	–	35	–	10	–	31	17	6
Brassica sp.	–	–	–	–	–	–	10	–	–	–	2
Unident. legume seed fragments	–	4	–	–	9	23	–	–	–	4	2
Total legumes	6	8	70	–	44	23	20	–	31	21	10
Amaranthus sp. (SFH)	3	1	9	–	–	–	–	–	–	3	5
Carex sp. (SFH)	3	2	–	–	17	–	73	–	–	8	6
Chenopodium sp. (SFH)	3	–	–	–	9	–	21	–	–	1	3
Cyperus sp. (SFL)	–	2	–	–	–	–	–	–	–	–	3
Dactyloctenium sp. (SFL)	–	–	–	–	–	–	–	–	–	–	–
Digera sp. (SFH)	3	2	–	15	–	–	–	–	–	3	2
Digiteria sp. (SFH)	3	–	9	–	35	–	–	–	–	6	–
Euphorbia sp. (SFH)	–	6	–	–	9	–	–	–	–	6	–
Mollugo sp. (SFH)	25	84	26	–	43	23	21	–	31	61	44
Phyllanthus sp. (SHH)	–	1	–	–	–	–	–	–	–	1	2
Solanum sp. (SFH)	–	2	9	–	9	–	10	–	–	4	2
Trianthema sp. (SFH)	9	7	–	–	9	45	–	–	–	6	10
Urochloa sp. (SFL)	3	43	–	–	9	–	42	–	–	3	58
Total Weeds	52	150	53	15	140	68	167	–	31	102	135

SFH = Small Free Heavy; SFL= Small Free Light; SHH = Small Headed Heavy

cies in samples from features and related specific contexts, instead they occur in higher frequencies in the general deposits (Table 5-2).

As an assemblage, weeds occur in the highest densities in Occupation I, followed by Occupation II (Table 5-2). Additionally the weeds at Oriyo Timbo occur in higher densities in the occupation surfaces associated with food preparation features. Among the weed species the dominant weed, *Mollugo* sp., occurs in highest densities within Occupation II. It is interesting to note that Occupation II stratum 5A, which had highest densities of millets and legume seed fragments, has relatively low frequencies of *Mollugo* sp. seeds. The implications of this distribution to the interpretation of the site's subsistence economy and also the role of *Mollugo* sp. will be addressed in the next chapter along with other related issues. Due to considerable bioturbation and other disturbance on Oriyo Timbo, further stratigraphic interpretations are limited; instead, emphasis is placed on the contextual relationships discerned from the data.

Contextual Relationships

The context of the samples is critical to their ultimate interpretation, and also to the interpretation of the archaeobotanical assemblage at the site. The context of the samples can be indicative of the use, initial discard, and status of plants in the archaeobotanical assemblage. At Oriyo Timbo, the samples are mostly from general sediment deposits within the occupation surfaces and fills not associated with specific contexts. However, there are some samples that come from definite contexts and features related to food preparation and dumping/discard. These contexts are positive indicators of human behavior specifically in relationship to the plant materials, and the following discussion on the contextual patterning of the archaeobotanical materials will stress these implications.

Trench S2 has no samples from features or related contexts, thus all the samples studied from this trench are those from the general sediment deposits associated with the different occupations and fills. It is however interesting to note that Trench S2 has the highest density of carbonized millets and legume seeds and lower densities for weeds. Among the millets recovered in Trench S2, *Setaria* spp. dominates the millet assemblage, with *Eleusine* sp. and *Panicum miliare* occurring in moderate densities. The legume category in the trench primarily includes *Phaseolus mungo*. The carbonized macrobotanical assemblage from this trench is thus dominated by crop plant seeds. The contexts of these seeds from Trench S2 (of general deposits/fills) are more indicative of an amalgam of different activities resulting from various events that lack resolution due to several factors. Included among these factors are the nature of the site, locations in which specified activities do occur, post depositional events which might have obscured these activities, and lastly repetition of activities within spatially confined areas (or the lack of such repetition in Trench S2).

The majority of samples from Trench S5 (80%) are from general deposits related to occupation surfaces and fills, although 20% are from specific feature's contexts related to food preparation and dumping/discard. In general, carbonized seeds were recovered in low densities from this trench. Millet seeds, represented only by *Panicum miliare* occur in lower densities, legumes (represented only by *Phaseolus mungo*) occur in moderate densities, while the weed seeds dominate the archaeobotanical assemblage of this trench.

In Trench S6 10% of the samples are from features, feature fills, and fills outside of features. Recovery of carbonized materials from this trench was moderate to high and it has the second highest density at the site. Millets dominate the archaeobotanical assemblage from S6, with weeds constituting a relatively small component, and legumes occur in low densities (Table 5-2). Among the millets, *Eleusine* sp. and *Setaria* spp. seeds dominate the assemblage, while *Phaseolus mungo* and *Brassica* sp. constitute the legume category. Trench S6 has the widest range of weed species, with different species in widely varying frequencies. Stratigraphically all the features in Trench S6 are in Occupations II and I. The samples from ash pits and food preparation features contain carbonized seeds of *Eleusine* sp., *Panicum miliare*, *Setaria* spp., *Carex* sp., *Mollugo* sp., and *Trianthema* sp. A clay-lined feature, however, had no carbonized seeds. The ashy pit contents were most probably cleanings and dumpings from living floors, including food preparation areas.

In Trench S8, 40% of the samples are from contexts related to food preparation and dumping/discard. The recovery of carbonized seeds is high regardless of context, and it has the highest density of carbonized seeds at the site. However, the frequencies of millet seeds and legumes are relatively low (Table 5-2). The archaeobotanical assemblage of the trench is dominated by weed species seeds (85%). *Mollugo* sp. is dominant, constituting over 90% of the weed species. *Mollugo* sp. most often grows as a weed in crop fields. The seeds are inedible, though the leaves could have been consumed. In Trench S8, samples from contexts of ashy pit features (which are most probably cleanings from the food preparation areas and living floors) and scrapings above burnt hearth surfaces contain a high percentage of *Mollugo* sp. seeds. It is very likely that the *Mollugo* sp. seeds were contaminants in the food grains and were selected out and ultimately found their way into the discard/dumping contexts that are found archaeologically. It is also likely that they were a part of the fuel used in the specific area (intentional or unintentional). Samples from defined contexts in Trench S8 vary in distributions of plants. A feature shaped like a keyhole (most likely a hearth) in Occupation II is sterile for carbonized seeds. Samples from features, ashpits, and fills above hearth/burnt surfaces are dominated by weeds while millets seeds show relatively lower densities in these contexts. The seeds include *Eleusine* sp., *Setaria tomentosa*, *Setaria italica*, legumes, *Digera* sp., *Cyperus* sp., *Solanum* sp., and *Urochloa* sp.

Trench S9 has the highest number of samples (81%) from defined features related to food processing and discard/dumping. In this trench, the samples in the fill separating Occupation II and III (stratum 4) are sterile for carbonized plant remains, while the samples from Occupation II have preserved carbonized seeds. The contexts of these samples are feature fills, and fills around, above, and below features. This includes a tandoor/oven wall with one carbonized seed of *Mollugo* sp., while the fill of the tandoor/oven has 3 carbonized seeds of *Mollugo* sp. and considerable charcoal. Sediments from three hearths in Trench S9 were also processed for plant remains. Of these, only one hearth has scanty plant materials preserved, comprising of one *Eleusine* sp. carbonized seed.

Contextual patterns in Oriyo Timbo carbonized macrobotanical data indicate that features not directly related to food preparation, such as hearth and oven contents, have a higher recovery of carbonized crop seeds such as those of *Eleusine* sp., *Setaria italica*, *Setaria tomentosa*, *Panicum miliare,* and *Phaseolus mungo* (Figure 5-3). These contexts include fills around, above, and below the hearths and a range of ashy features. The implications of this patterning are significant. Thus, it is argued that Oriyo Timbo carbonized plant remains appear to be differentially preserved based on context (Reddy 1991b). Contexts with the highest probability of fire and high temperatures, such as hearth contents, tandoor/oven fills and walls, all appear to have very low densities of carbonized seeds. Whether this correlation is a result of a lack of preservation due to the high temperatures in these contexts, or a result of no seeds being associated or

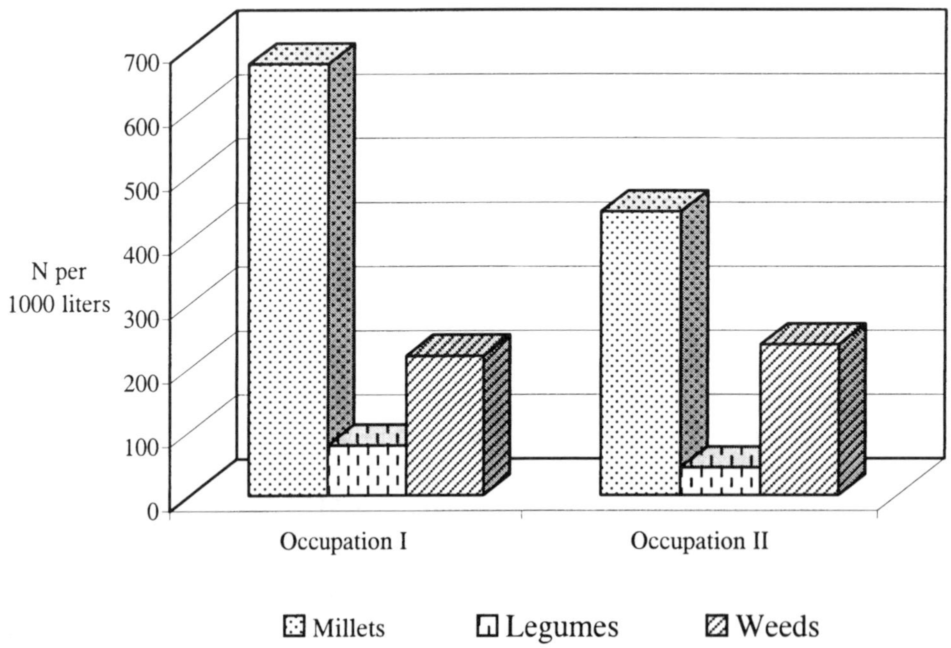

Fig. 5-2. Oriyo Timbo: Occupational distribution of millets, legumes and weeds.

incorporated into these contexts is a challenging question. It is asserted here that there is a very high probability of seeds accidentally getting incorporated into hearths during food preparation, but the high temperatures of the hearths would have reduced them to ash very soon after, therefore their survival would be minimal. Ethnographic observations revealed that accidental spillage of seeds during food preparation and also during the cleaning process before food preparation is very likely and occurs in a significant number of incidents. Thus, for Oriyo Timbo, it is argued that the significantly low densities of carbonized seeds from hearths are due to their being destroyed as a result of the high temperatures rather than not being initially incorporated into this context.

It has also been noted at Oriyo Timbo that the highest seed densities occur in samples from deposits around, above and below features (Figure 5-3). This is due to depositional processes during and after these activities. For example, in India, most final cleaning of food grains is done nearby or around the hearth (almost as an after thought) most often directly before the food preparation. Sometimes the cleaning occurs in the front porches/yards. In either case, the cleanings are most often swept into the hearth after the cooking is done, that is while the fire is dying or after the fire is put out. This provides the ideal temperature for the seeds to be carbonized but not destroyed. Subsequent cooking, however, might destroy these seeds unless the hearth was abandoned soon after such an event (which is what we would then discover archaeologically), or unless the hearth was cleaned out and the contents disposed into a trash/dump area. In either case, it is important to note that intentional and unintentional discard of crop grains occur in these contexts due either to their immaturity, other undesirable quality, or accidental disposal.

Miller and Smart (1984) have suggested that a significantly high percentage of archaeobotanical materials retrieved from living floors, hearths, and ovens are preserved as a result of dung being intentionally burnt as fuel, and also due to dung plastering of floors. However, it is important not to subsequently assume that the archaeobotanical materials from these contexts specified by Miller and Smart (1984) are related to animal fodder. It is equally probable that the grains included in dung could come from elsewhere (i.e., inclusions when the dung was dropped, dried, packed, etc.), rather than just from the animal's digestive system. These variables can only be controlled and explained upon thorough understanding of the context of the archaeological sediment. Alternatively, dung need not be the primary source of seeds in archaeological sediments. Crop byproducts are also often used as fuel in the Third World, although millet byproducts rank quite low as attractive domestic fuels when compared to wheat, barley, and sugarcane crop byproducts (Barnard and Kristoferson

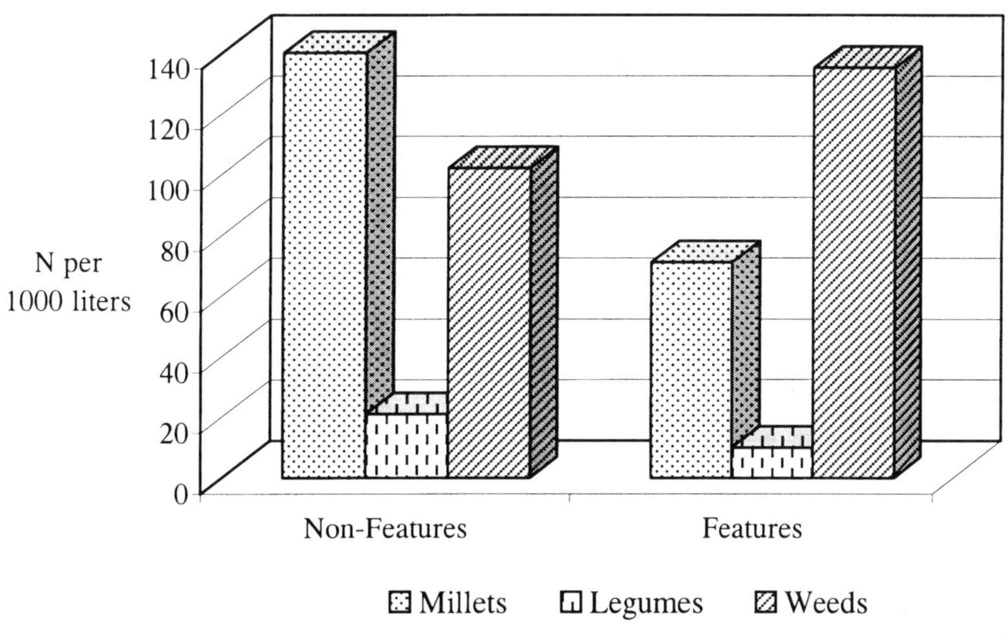

Fig. 5-3. Oriyo Timbo: Contextual patterns of millets, legumes and weeds.

1989:79). So how can one understand the contextual distribution of the carbonized seeds recovered at Oriyo Timbo? I propose that most of the crop seeds recovered at Oriyo Timbo are not from dung being used as fuel, but rather their inclusion in the archaeological record is evidence of their use as food. This hypothesis is tested through the application of ethnographic models.

Summary of Oriyo Timbo

To summarize the Oriyo Timbo 1989-90 archaeobotanical results, there is strong evidence for the use of summer crops. There is patterned variation in the frequencies of carbonized seeds recovered between trenches. Trench S8 has the highest density of weed seeds, which accounts for it having the highest carbonized seed density, while Trench S2 has the highest millet seed density. Stratigraphically Occupation II and I have higher seed densities as compared to the fills between the occupations (Figure 5-2). There is a significant difference between the densities of carbonized seeds recovered from Occupation II and I, with Occupation II having a significantly higher carbonized seed density. Given this pattern coupled with the higher number of food preparation features in Occupation II, it is argued that at Oriyo Timbo there is an increase in crop plant seed usage over time, that is from Occupation I to Occupation II.

It is also clear that the distribution of carbonized seeds at the site is not concentrated in any functional locus, such as a hearth or oven. In fact, the samples from these contexts show a relatively low density of seeds compared to the general sediments from the different trenches (Figure 5-3). This may be due to several factors, the most probable being that the preservation of seeds in these contexts is lower due to the high temperatures of cooking and burning, which would totally, destroy the archaeobotanical materials. There are several important questions that need to be answered to fully understand the Oriyo Timbo archaeobotanical assemblage and subsequently interpret the subsistence economy at the site. These questions, as already discussed, include whether the summer crops such as *Eleusine* sp., *Setaria* spp., *Panicum miliare* and *Phaseolus mungo* were cultivated at the site or nearby, or were brought as grain from other locals? Equally important is the need to elucidate the economic importance of these crops to the Oriyo Timbo inhabitants. If the crops were being cultivated, were the millets used for primarily food for humans and fodder for animals?

Babar Kot Archaeobotanical Results

The main crops represented at Babar Kot include 'millet' crops (such as *Setaria* spp., *Panicum* sp., and *Eleusine* sp.), *Linum* sp., and several legumes. The objective of this archaeobotanical study is to present a working model for the economic importance of the millet crops (such as *Setaria* spp., *Panicum* sp., and *Eleusine* sp.) in the subsistence system at Babar Kot, and to determine whether these crops were being cultivated by the site occupants. During the 1990-1991 excavation season at Babar Kot both carbonized and uncarbonized seeds were recovered from the excavation trenches. The sediment samples from Babar Kot averaged 7.8 liters in volume, and the archaeobotanical recovery averaged 18.25 carbonized seeds per liter of soil sediment. The high density of seeds at the site is primarily associated with burnt soil rather than with high quantities of charcoal. This association is similar to the pattern of archaeobotanical recovery at the site of Oriyo Timbo where the contexts with high quantities of charcoal appeared to have the lowest frequency of carbonized seeds (Reddy 1991b:81).

The plant remains occur in varying densities across the excavated trenches. For the purpose of the discussion, the excavation areas will be referred to as mound units and slope units. The recovery of carbonized archaeobotanical materials is higher in the slope units, as compared to the mound units (Table 5-3). However, the proportions of millets to legumes to weeds are similar in both the slope and the mound units. There is patterned variation in the stratigraphic distribution of the carbonized seeds in general and also the groups of plants (i.e., millets, legumes, and weeds). Occupation III shows the highest density of carbonized seeds recovered, while Occupations II and I show a lower but similar recovery (Table 5-4).

The crop seeds *Setaria* spp. and *Panicum* spp. occur in relatively high densities at the site, while those of *Linum* sp., *Eleusine* sp., *Lathyrus* sp., *Lens* sp., and *Vicia* sp. occur in lesser amounts. Most of these crops are summer-cultivated plants. A variety of non-crop seeds were also recovered including those of *Aizoon* sp., *Carex* sp., *Chenopodium*/*Amranthus* sp., *Cyperus* sp., *Digitaria* sp., *Echinochloa* sp., *Eragrostis* sp., *Ficus* sp., *Lotus* sp., *Medicago* sp., *Phyllanthus* sp., *Polygonum* sp., *Solanum* sp., *Trianthema* sp., *Urochloa* sp., and *Zizyphus* sp. Of the total assemblage at Babar Kot, summer cultivars comprise 86%, legume crop/winter cultivar seeds comprise 0.4% and non-crop seeds

Table 5-3. Babar Kot. Carbonized Seed Densities on the Mound and the Slope.

	26L	26M	Mound Total	36M	38P	40M	40N	Slope Total
Sediment Floated (liters)	285	52	337	6	129	62	29	226
Carbonized Seeds (Absolute Counts)	1439	255	1694	11	15552	168	1094	16825
Seed Densities N per 100 liters	**26L**	**26M**	**Mound Total**	**36M**	**38P**	**40M**	**40N**	**Slope Total**
Carbonized Seeds	505	490	503	183	12053	271	3772	7445
Unidentified Seeds	114	-	97	-	488	32	383	336
Big Seed fragments	1	-	0.60	-	6	-	-	4
Nut Shells	-	-	-	-	1	-	-	1
Eleusine coracana	2	4	2	-	-	-	-	-
Panicum spp. (total)	152	4	129	-	559	24	190	350
Panicum miliaceum	-	-	-	-	-	-	14	2
Panicum miliare	152	4	129	-	559	24	176	348
Pennisetum sp.	-	-	-	-	1	-	-	0.44
Setaria spp. (total)	160	458	206	183	10594	60	1907	6315
Setaria glauca	2	10	3	-	-	-	-	-
Setaria italica	157	448	202	183	10347	58	1907	6172
Setaria viridis	1	-	0.60	-	-	2	-	0.44
Setaria spikelets with rachillae	-	-	-	-	247	-	-	142
Cereal Rachii	3	4	3	-	22	-	10	14
Total Carbonized millets seeds	317	470	340	183	11176	84	2107	6679
Linum sp.	-	-	-	-	5	2	86	15
Brassica sp.	-	-	-	-	-	-	3	0.44
Lathyrus sp.	0.35	-	0.30	-	-	-	-	-
Lens sp.	4	-	4	-	1	3	10	3
Vicia sp.	-	-	-	-	-	10	3	3
Total legumes	4.35	-	4	-	1	13	16	6
Aizoon sp. (SFH)	1	-	1	-	1	-	-	0.44
Carex sp. (SFH)	2	-	2	-	1	2	-	1
Chenopodium sp. (SFH)	0.35	-	0.30	-	1	-	3	1
Cyperus sp. (SFL)	41	-	35	-	43	-	62	32
Digitaria sp. (SFH)	-	10	2	-	60	42	348	91
Echinochloa sp. (SFH)	0.35	-	0.30	-	1	-	41	6
Eragrastis sp. (SFL)	-	-	-	-	134	-	-	76
Ficus sp. (SFH)	-	-	-	-	-	2	-	0.44
Lotus sp. (SFH)	-	-	-	-	2	2	-	1
Medicago sp. (SFH)	-	-	-	-	-	2	3	1
Phalaris sp. (SFL)	1	-	1	-	-	-	-	-
Phyllanthus sp. (SHH)	0.35	-	0.30	-	-	-	21	3
Polygonum sp. (SFL)	4	-	3	-	-	-	-	-
Solanum sp. (SFH)	1	2	1	-	-	-	-	-
Trianthema sp. (SFH)	19	10	17	-	9	34	-	15
Urochloa sp. (SFL)	-	-	-	-	123	58	697	175
Ziziphus sp. (BFH)	-	-	-	-	5	-	-	3
Total Weeds	70.05	22	62	-	376	142	1178	405

Table 5-4. Babar Kot. Carbonized Seed Densities by Occupation and Context.

N per 100 Liters	**OCCUPATIONS**			**MOUND**		**SLOPE**	
	III	**II**	**I**	**Non-Features**	**Features**	**Non-Features**	**Features**
Carbonized Seeds	7539	707	715	350	1251	1625	11718
Unidentified Seeds	178	164	110	79	184	191	436
Big Seed fragments	2	1	–	1	–	–	6
Nut Shells	1	–	–	–	–	–	1
Eleusine coracana	1	1	–	3	–	–	–
Panicum spp. (total)	357	94	410	65	440	92	527
Panicum miliaceum	1	–	–	–	–	–	3
Panicum miliare	356	94	410	65	440	92	524
Pennisetum sp.	0.3	–	–	–	–	–	1
Setaria spp. (total)	6625	393	–	156	449	1151	9860
Setaria glauca	1	3	–	4	–	–	–
Setaria italica	6473	329	–	151	449	1151	9625
Setaria viridis	2	–	–	1	–	–	1
Setaria spikelets with rachillae	150	7	–	–	–	10	234
Cereal Rachii	1	4	–	3	4	–	23
Total Carbonized millets seeds	6998	492	413	227	893	1243	10410
Linum sp.	10	1	–	–	–	3	22
Brassica sp.	0.3	–	–	–	–	–	1
Lathyrus sp.	–	0.3	–	–	2	–	–
Lens sp.	2	0.3	25	4	4	–	4
Vicia sp.	2	–	–	–	–	–	5
Total legumes	4	1	25	4	6	–	10
Aizoon sp. (SFH)	–	–	10	1	2	–	1
Carex sp. (SFH)	1	1	–	2	–	–	1
Chenopodium sp. (SFH)	1	1	–	0.3	–	–	1
Cyperus sp. (SFL)	31	24	88	16	126	–	54
Digitaria sp. (SFH)	71	–	–	1	4	22	138
Echinochloa sp. (SFH)	3	0.3	–	–	2	–	10
Eragrastis sp. (SFL)	81	–	–	–	–	–	128
Ficus sp. (SFH)	0.3	–	–	–	–	–	1
Lotus sp. (SFH)	1	–	–	–	–	2	1
Medicago sp. (SFH)	1	–	–	–	–	–	1
Phalaris sp. (SFL)	–	1	–	1	–	–	–
Phyllanthus sp. (SHH)	2	1	–	0.3	–	–	4
Polygonum sp. (SFL)	2	4	–	4	–	–	–
Solanum sp. (SFH)	0.3	1	–	1	–	–	–
Trianthema sp. (SFH)	10	10	69	14	35	2	23
Urochloa sp. (SFL)	138	6	–	–	–	48	263
Ziziphus sp. (BFH)	3	–	–	–	–	–	5
Total Weeds	346	49	167	41	169	74	631

comprise 6%. Among the crops, *Setaria* spp. seeds are by far the most prevalent (92%), while *Lens* sp. is the most frequent among the legumes (67%). *Urochloa* seeds account for the most ubiquitous non-crop plant (48.6%) (Table 5-4).

Millets

A total of 16,198 carbonized seeds belonging to four genera, *Eleucine* sp. (n=8), *Panicum* sp. (n=1,225) *Setaria* spp. (14,964), and *Pennisetum* sp. (n=1) commonly referred to as millets was recovered from Babar Kot (Table 5-3). In addition, 41 rachis remains were also recovered. *Eleusine coracana* was recovered exclusively from the mound units. *Panicum* sp. constitutes 7.5% of the millet assemblage, and includes 1,221 seeds of *Panicum miliare* and 4 seeds of *Panicum miliaceum*. The *Setaria* spp. seeds include three genera accounting for the highest seed density of all carbonized seeds recovered from Babar Kot; *Setaria glauca* (n=11), *Setaria tomentosa* (n=3), *Setaria italica* (n=14,628), and *Setaria italica* spikelets with rachillas (n=322). It is proposed that the primary use of *Setaria italica* at Babar Kot is as human food, though secondary use as fodder, specifically of the crop's byproducts, is quite likely. Of the 14,628 seeds of *Setaria italica*, 8,715 seeds were recovered from a seed pocket. One seed has been tentatively identified as belonging to the genus *Pennisetum*. It was recovered from a slope unit in a feature fill associated with Occupation III. Recovery of the *Pennisetum* sp. seed is very significant however, considering its singular presence no further discussion is warranted. (Please refer to Reddy [1994] for further details.)

Cereal Rachis Fragments

A total of 41 rachis fragments were recovered at Babar Kot, of which 26 were from the seed pocket discussed above (Table 5-3). These rachii have been identified as those of *Setaria* sp. The cereal rachii are indicative of crop processing activities, and aid in defining activities related to domestic food processing. In addition, they can be used to model the role and status of *Setaria italica* and millets in general at Babar Kot.

*Oilseeds (*Linum *sp.)*

A total of 33 seeds were recovered and identified as an oilseed (Table 5-3). Initially they were tentatively identified as *Sesamum* sp. but further detailed analysis invalidated that identification, and they are confirmed as belonging to *Linum* sp. The seeds are identified as *Linum* sp., but their species is not certain. The oilseed can be grown as a cold or warm weather crop depending on the species. The crop is grown for the seed, and specifically to extract oil from it. Traditionally it was also grown as animal fodder.

The presence of *Linum* sp. seeds is important because it demonstrates that the inhabitants of Babar Kot were not totally concentrating their subsistence base on millet crops, but the oilseed was also part of their subsistence system. Whether they were extracting oil from the seeds or not cannot be conclusively demonstrated, however, the condition of the recovered oilseeds sheds some light on this issue. All the 33 oilseeds showed signs of rupture on the narrow end, which could be indicative of their being crushed for oil. This must remain speculation until ethnographic studies can confirm such morphological effects, or other patterns indicative of crushing these seeds for oil. However, it does remain an intriguing speculation with important implications on the issue of the breadth of the subsistence base at Babar Kot.

Legumes

A total of 27 carbonized seeds have been identified as being leguminous, and they belong to the genera *Brassica* sp. (n=1), *Lathyrus* sp. (n=1), *Lens* sp. (n=18), and *Vicia* sp. (n=7) (Table 5-3). They occur in varying frequencies, and also demonstrate varying spatial and temporal distribution. Their presence further demonstrates that the Babar Kot inhabitants had a relatively broad subsistence base. In addition, the presence of *Lathyrus* sp. and *Lens* sp. indicate winter cultivation/procurement, since these crops are most often grown in cold weather.

Non-Crop Plants

A total of 17 different non-crop plant genera, termed weeds, were recovered from the site (Table 5-3). Among all these non-crop seeds, *Urochloa* sp. occurs in the highest densities, and constitutes 35% of the weed assemblage. The other weeds that occur in relatively high proportions are *Cyperus* sp. (17%), *Digitaria* sp. (19%), *Eragrostis* sp., and *Trianthema* sp. (8%). The 17 different genera can be classified into four different weed categories, which include small free light (SFL) category (69%), small free heavy (SFH) category (30%), small-headed heavy (SHH) category (0.6%), and big free heavy (BFH) category (0.6%). It is important to note that not all the weed categories are represented. The ethnographic modeling has demonstrated that specific categories of weeds can be used as indicators of different stages of crop processing, and the categories present in the Babar Kot weed assemblage (SFL, SFH, SHH, and BFH) will be used to shed light on what the archaeobotanical assemblage at the site is most indicative of; earlier processing stages or later processing stages. This information can be used to infer whether the crops were cultivated at the site or brought in through exchange from elsewhere. Weeds will be also used to identify activities and activity areas related to crop processing and food processing, since different characteristics of the weeds can be used to identify different stages of processing (Reddy 1991b).

Spatial Trends

Comparing the archaeobotanical data from the two excavations areas at the site, the slope and the mound, it is noted that there is a higher recovery of carbonized plant remains from the slope units per liter of sediment floated (Table 5-3). In other words, the density of carbonized seeds is significantly higher on the slope, and this is reflected in higher densities for summer cultivars/millets, legumes, and weeds. In both areas, summer cultivars have the highest densities followed by non-crops, and then by legumes. The higher densities of carbonized seeds in the slope could be related to context. There are more feature context samples from the slope units, and there may be a direct correlation between the context of the sample and the density of seeds recovered.

There is some spatial patterning in the recovery of particular genera from the slope and the mound (Table 5-3). *Lens* sp. seeds occur in higher densities in the slope and exclusively in feature contexts. *Panicum miliare* caryopsis occur in higher densities in the slope, while *Eleusine* sp. and *Lathyrus* sp. were recovered only from mound units. *Linum* sp., *Brassica* sp., and *Vicia* sp. were recovered only from the slope. Of these, both *Brassica* sp. and *Vicia* sp. occur in relatively low frequencies. *Setaria* spp. occur in significantly higher densities in the slope as compared to the mound. Of significance is the recovery of *Setaria* sp. spikelets with rachillas and *Setaria* sp. rachii from only the slope, and their prominent absence in the mound. These two components are used to identify crop-processing activities, since they are processed out at specific stages. Their explicit presence in the slope but not in the mound unit, leads to unmistakable deductions regarding the spatial differentiation of various activities. The crop processing activities related to the selection/cleaning out of *Setaria* sp. spikelets with rachillas and the *Setaria* sp. rachii were being performed primarily in the slope areas during Occupation II and III (Table 5-3 and 5-4).

Another important spatial distinction occurs within the weed assemblage (Table 5-3). The weed assemblage from the mound is characterized by the dominance of SFL weeds and SFH weeds, and a comparatively low frequency of SHH weeds. The slope weed assemblage, however, is primarily dominated by the SFH weeds, with SFH weeds occurring in significantly lower frequencies. In addition, the slope assemblage also has SHH weeds and BFH weeds in low frequencies but still more common than in the mound units. Thus, the slope units are dominated by SFL weeds along with crop plant spikelets with rachillas and rachii fragments. These associations and patterns are critical for the interpretation of crop processing activities, since different categories of weeds are processed out through specific processing stages. This information, coupled with the data on the distributions of *Setaria* sp. spikelets with rachillas and *Setaria* sp. rachii, is instrumental for the identification of specific processing activities.

Stratigraphic Patterns

There is considerable variation in the distributions of archaeobotanical remains recovered from the different occupational levels (Table 5-4, Figure 5-4). Occupation III has the significantly highest densities of carbonized seeds, and Occupations II and I have lower and comparable densities. Occupation III also has the most number of feature context samples. Occupation II has relatively few samples from feature contexts, while Occupation I has none.

Occupation I has the highest densities of legumes, and the second highest frequency of weed seeds (Table 5-4). The legume assemblage of Occupation I is exclusively represented by *Lens* sp., and the millet assemblage is represented only by *Panicum miliare*. The weed assemblage is represented equally by SFH and SFL weeds. Occupation II has the lowest density of carbonized seeds and weed seeds at the site (Figure 5-4, Table 5-4). The millet assemblage is represented by two crops, *Setaria* sp. and *Panicum miliare*. *Setaria* sp. spikelets with rachillas and rachii occur in very low densities in

Occupation II. The weed assemblage of Occupation II is dominated by SFL weeds, although SFH weeds are also well represented. Occupation III has the highest densities of carbonized seeds in all the groups except legumes (Figure 5-4, Table 5-4). This last occupation has the significantly highest densities of millets and weed seeds. Among the millets, *Setaria* spp. dominate the assemblage, specifically *Setaria italica*. Rachii fragments and spikelets with rachillas of *Setaria* sp. are also recovered in relatively higher densities. Legumes show a low density, although *Linum* sp. was recovered primarily in Occupation III. The oilseeds are primarily associated with feature contexts, specifically in an Occupation III pit, which also had animal bones and domestic refuse, other crop seeds, weed seeds, and rachii. The weed assemblage of Occupation III is dominated by SFL weeds, although SFH weeds are also moderately represented. In addition, the Occupation III weed assemblage also has SHH and BFH weeds in low frequencies. There is clear change in the ratios of the different weed assemblage categories from Occupation I through III. Stratigraphically, Occupation III has a significantly higher density (Table 5-4) of non-crop plant seeds.

Stratigraphically, *Setaria* spp. occur consistently in Occupations II and III, and overall they appear to be the most intensively used and cultivated crop at the site (Table 5-4). *Setaria* spp. are however absent in Occupation I, appearing first at the onset of Occupation II, and in very high densities in Occupation III. *Setaria* spp. spikelets with rachillas are present only in Occupation III. In addition, *Setaria* sp. rachii occur in significantly higher densities in Occupation III, in lesser densities in Occupation II and are absent in Occupation I. These rachis remains and spikelets with rachillas are key to establishing the status of the crop; in other words the presence of these components (in conjunction with other components such as the different weed categories) are indicators of specific crop processing activities which can shed light on whether the crop was cultivated by the inhabitants or traded in from elsewhere.

Eleusine coracana occurs only in stratum 4 of Occupation III and stratum 8 of Occupation II (Table 5-4). These occupation deposits relate to an amalgam of activities associated with human occupation and subsistence, and not to a specific activity. Stratigraphically, *Panicum* sp. varies between the two excavated areas with the lower strata of the mound units having higher densities of *Panicum miliare*, while the upper strata of the slope units have lower densities. *Panicum miliare* shows a differential distribution across the three occupations, with the millet assemblage of Occupation I being comprised only of *Panicum miliare* seeds. Occupation III has a similar density, but Occupation II has significantly lower densities of *Panicum miliare* seeds (Table 5-4). The lower densities of *Panicum miliare* in Occupation II suggest a lower usage of the crop during this occupation. However, the usage of the crop is primarily as

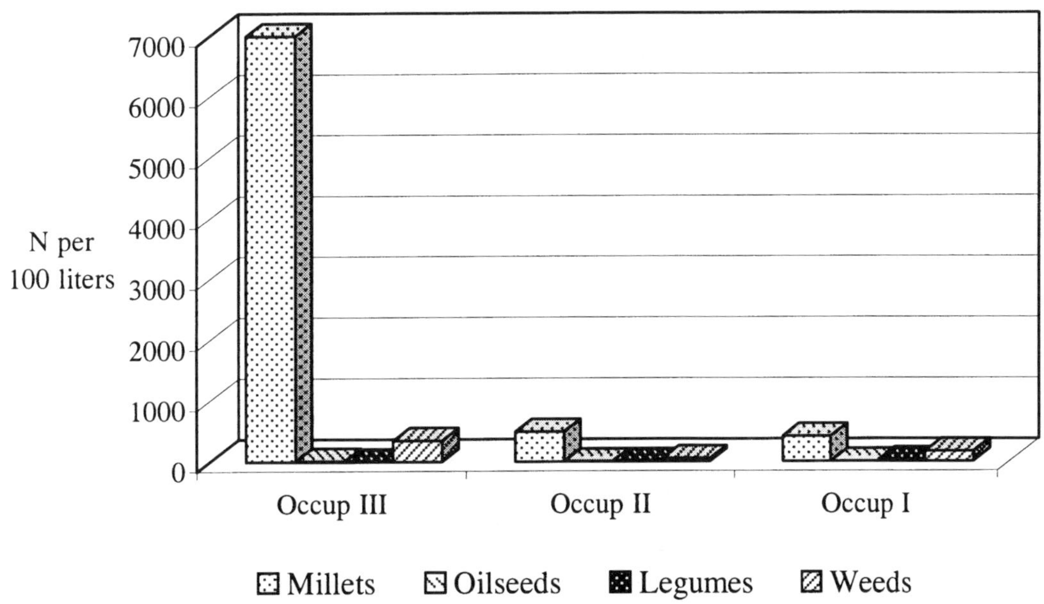

Fig. 5-4. Babar Kot: Occupational distribution of millets, oilseeds, legumes and weeds.

human food since a significant percent of the *Panicum miliare* in Occupation II comes from feature contexts, particularly dump pits and a clay feature (related to food preparation) (Table 5-4). *Panicum miliare* is considered to be an important crop plant in the Babar Kot assemblage. It is definitely not the dominant millet crop, but it was certainly heavily utilized during Occupations III and I (Table 5-4). Its decreased usage during Occupation II can be explained best by correlating it to the addition of *Setaria italica* into the millet assemblage during this occupation.

Interestingly, the relative frequency of *Setaria* spp. varies inversely with the relative frequency of *Panicum* sp. (Table 5-4). *Panicum* sp. occurs in highest densities in Occupation I and in much lower densities in the later occupations (with the exception of strata 5 and 5a of Occupation III). *Setaria* spp. are, however, absent in Occupation I and increase in Occupation II to III. This pattern may suggest that *Setaria* spp. are becoming more important as a food crop at the site at the expense of *Panicum* sp. This inference (that changes in the composition of chronologically different assemblages indicate changes in plant resource use) has its limitations since it assumes that the different temporal assemblages were all created by similar processing stages. Unless all the assemblages were recovered from storage facilities, this assumption may not be valid. A much stronger inference regarding changes in plant resource use over time must first examine whether the different assemblages, from various periods, are the result of similar processing stages or not. The ability to correlate particular archaeobotanical assemblages to certain stages in plant processing with some degree of certainly is crucial. Therefore, the nature of the *Panicum* sp. and *Setaria* spp. assemblages as they correlate to processing stages in the different occupations will be addressed in Chapter 6, specifically whether different processing stages and activities are responsible for the inversed frequencies of the two crop seeds in the different occupations.

Contextual Relationships

Contextual inferences are significant because they shed light on the initial status of the plant material; in other words, the motive for their discard and primary use can be inferred from the context of their discovery. At Babar Kot there is distinctive contextual patterning of archaeobotanical remains. Feature contexts of the Babar Kot samples include the seed pocket, ash pits, burnt surfaces, distinctive features, and pot contents. An important pattern discerned at Babar Kot is that the samples from defined feature contexts have a significantly higher density of carbonized seeds in contrast to samples from general fill sediments (Figure 5-5). This trend is distinct from the patterns observed at Oriyo Timbo where samples from feature contexts with high temperatures (such as hearth samples and tandoor fill contents) had the lowest carbonized seed density, while samples from deposits above and below features were noted to have the highest density of carbonized seeds. Additionally, the high density of seeds at Babar Kot is primarily associated with burnt soil rather than with high quantities of charcoal. This association is similar to the pattern of archaeobotanical recovery at the site of Oriyo Timbo where the contexts with high quantities of charcoal appeared to have the lowest frequency of carbonized seeds (Pearsall 1989; Reddy 1991b:81). These variations and similarities demonstrate that the collection of sediment samples for flotation should be systematic at all sites and contexts, and in addition reiterate the need for the paleoethnobotanist to participate in the excavations.

The densities of carbonized seeds recovered from the slope units are in general (regardless of contextual distinctness) significantly higher than the carbonized seed densities on the mound (Table 5-3, Figure 5-5). One of the contributing factors for this spatial discrimination is context, that is, distinct features caused by human behavior and activities. Two patterns in the data support this argument. The first relates to the higher densities of carbonized seeds in specific feature context samples. In addition, there are more feature context samples from the slope units. Thus, given these two observations it is argued that there is a direct correlation between the context of the sample and the density of seeds recovered. In other words, the intentional discard of seeds into features is occurring in greater frequencies than the incorporation of seeds into the general occupation deposits.

The densities of features clearly vary stratigraphically between the three occupations. Occupation III has the most number of features, accounting for 74% of the total features excavated. Occupation II has 26% of the features excavated, while Occupation I contained no features. It is very important to note that the densities of carbonized seeds recovered also follow this succession, with Occupation III having the significantly highest density of recovered carbonized seeds (Table 5-4, Figure 5-4).

Different plant species recovered at the site appear to be differentially associated with specific feature contexts (Table 5-4, Figure 5-5). Among the millets there is a definite variation in contextual recovery. For example, *Eleusine coracana* is differentially correlated to context, occurring specifically in the non-feature contexts (occupation deposits not clearly associated with any particular activity such as food preparation). This millet was recovered only from the mound unit, and specifically in non-feature contexts (Table 5-3). The contexts of *Eleusine* sp. recovery relate to an amalgam of activities related to human occupation and subsistence, and not to a specific activity. What this suggests is that *Eleusine* sp. cannot be explicitly linked to human usage. In contrast, the recovery of *Panicum miliare* seeds is significantly higher in feature contexts, which indicates that it was primarily used as human food. Typically *Setaria italica* and *Panicum miliare* occur in significantly higher densities in feature contexts than in general deposits, thus linking them unquestionably to specific human activities (Table 5-4). For example, about 90% of the *Setaria italica* seeds were recovered from feature contexts in the slope units, and about 75% of the *Setaria italica* seeds were recovered from the feature contexts in the mounds. The *Setaria italica* spikelets with rachillas also occur in significantly higher densities in feature contexts. The *Setaria* sp. rachii occur in significantly higher densities in features contexts. In the mound units, the rachii occur in lower density, and in equal densities in feature and non-feature contexts. This spatial distinction could be indicative of differential activity areas at the site, specifically the slope areas being the primary locales for crop and food processing. Regardless of this spatial variation within the site, their significantly higher densities in feature contexts link them unquestionably to specific activities related to human behavior pertaining to crop processing for human food. The features are a direct link to these activities, providing an inverse picture of the activities, which can be reconstructed using ethnographic data. *Setaria glauca* in contrast occurs only in non-feature contexts and only in the mound units. *Setaria viridis* does not exhibit any distinctive pattern of distribution. Both of these species, *Setaria glauca* and *Setaria viridis*

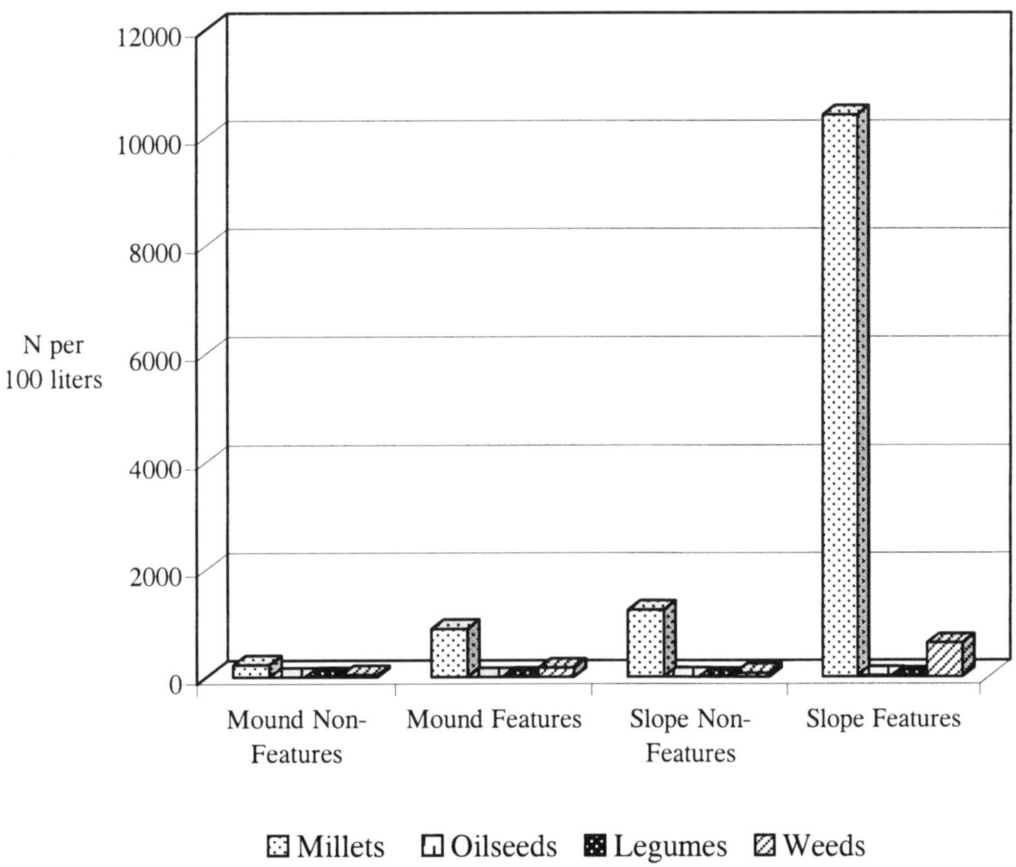

Fig. 5-5. Babar Kot: Contextual patterns of millets, oilseeds, legumes and weeds.

are not considered to have been used as human food, and their presence in the assemblage is attributed to other factors.

Contextually, the five major weed genera (*Urochloa* sp., *Cyperus* sp., *Digitaria* sp., *Eragrostis* sp., and *Trianthema* sp.) occur in significantly higher densities in feature contexts, with *Eragrostis* sp. occurring exclusively in feature contexts (Table 5-4, Figure 5-5). This pattern indicates that the presence of *Eragrostis* sp. seeds is a result of a specific human activity and they are not non-cultural intrusives. The different weed categories occur differentially based on context, suggesting that there is a relationship between context and the weed category. The weeds of the BFH and SHH categories occur exclusively in feature contexts, thus connecting them to human activities. Weeds of SFL and SFH categories occur in both feature and general occupation deposits, but they occur in significantly higher densities in feature contexts. The implications of this patterning are explicit in that it is highly probable that the presence of the weeds is a result of specific human behavior and activities related to crop processing and food preparation.

A total of seven features was encountered and excavated at Babar Kot which were all sampled for archaeobotanical remains and yielded high densities of carbonized seeds. The most impressive feature recovered at Babar Kot is an Occupation III seed pocket of *Setaria* sp. from the slope. Other features include an Occupation III trash pit on the slope, Occupation III Harappan pot contents on the mound, an Occupation III refuse ash pit on the slope, two Occupation II pits on the mound, and an Occupation II clay feature on the mound.

Occupation III Seed Pocket, Slope

Of the 14,628 seeds of*Setaria italica*, 8,715 seeds were recovered from a seed pocket in Occupation III (stratum 5) of the slope units. It is important to emphasize here that *Setaria* spp. seeds occur in significantly higher densities in the slope units with or without the inclusion of the seed pocket seeds. The recovery of this seed pocket or lens with carbonized *Setaria* sp. seeds is of great significance. The seed pocket is predominantly composed of *Setaria italica* caroypsis (85%), rachii (0.3%), and spikelets with rachillas (3%). An additional 6% of the seeds recovered from this feature are *Panicum miliare*, and 3% are comprised of *Cyperus* sp. seeds, *Digitaria* sp. seeds, *Eragrostis* sp. seeds, *Urochloa* sp. seeds, and *Zizyphus* seed/nut fragments. Thus, a significant percentage of the seed pocket is of *Setaria italica* caryopsis. This archaeobotanical find at Babar Kot is particularly important because to date there have been no similar finds in Gujarat, and the high number of carbonized *Setaria* spp. seeds which are mostly clumped together in this feature is intriguing. There is no associated disturbance, and there are signs of burnt soil and ash in the vicinity. The seed pocket was found immediately below a distinct feature of loose, ashy, fine sediment defined as a small pit. Since the trench that contained this feature was excavated by the author, it can be strongly asserted that no insect, rodent or other animal disturbances were noted. The well-carbonized state of the seeds negates the possibility of insect-hole activity, and the clumped nature of the seeds through burning negates the role of disturbance or animal activity. The associated contexts were those of domestic refuse and domestic dumping, with ash pits and domestic discard dominating the unit. Given this situation, it is safe to state that this seed pocket is a direct result of human activity related to food preparation. In this light the burnt and clumped nature of the seeds are very critical to their interpretations and will be addressed in Chapter 6.

Occupation III Trash Pit, Slope

A domestic refuse pit was encountered in a unit on the slope and excavated in four levels, and there is a distinct pattern to the archaeobotanical remains recovered from the different levels. This particular feature pit has a relatively high density of carbonized seeds (1094 carbonized seeds in 28.5 liters of sediment), along with animal bones and other domestic refuse such as burned ceramics, clay lumps, and ash. The general sediment of the pit was burnt and ashy, and the pit is interpreted as being used as a trash dump with a series of disposal episodes. Carbonized seeds recovered from this dump/trash pit include millets seeds (*Panicum* sp., *Setaria* sp., and *Setaria* rachis fragments), legumes (*Lens* sp., *Vicia* sp., and *Brassica* sp.), and a range of weeds. Among the weeds, the category significantly represented is that of the SFL, accounting for 64% of the weed assemblage. SFH weed seeds occur in lower frequencies accounting for 34% of the weed assemblage in the pit, while SHH weeds account for only 2% of the assemblage.

There appeared to be a stratigraphic patterning in the carbonized seeds recovered from this feature. The bottom level of the pit has a low density of carbonized seeds (0.2% of the total carbonized seed assemblage of the pit) with only 2 seeds of *Setaria italica*. The middle two layers of the pit accounted for the majority of the assemblage with 79% of the seeds. The assemblage of these middle layers is mainly dominated by *Setaria* sp., although *Urochloa* sp. also occurs in relatively high densities. *Linum* sp. occurs in significantly higher densities in these middle levels, as do the *Setaria* sp. rachii. The top level of the pit is characterized by the presence of *Panicum* sp. and *Setaria* sp. in equal proportions. Legumes occur exclusively in this upper layer of the pit.

The stratigraphic variation of the different plant remains in the pit is very significant, because it is indicative of sequential dumping episodes, and also the differences in the dumping contents during each episode. The argument for sequential dumping (as opposed to one episode) is based on two lines of evidence. First, the relatively deep pit stratigraphy reveals a sequence of distinct ashy gray layers representing different episodes of dumping. Second and equally important is the wide range of plant remains recovered from the pit. If it was a single episode of dumping, then it is more likely to have a very limited range of genera/species, perhaps only one or two, since different plants are not processed for food at the same time and mixing of the genera typically would not occur. Thus, the wide range of plants recovered, the stratigraphic variation in the pit, and the patterned variation in the distribution of different species between levels are all indicative of a series of dumping episodes.

Based on the patterning of the recovered archaeobotanical remains in the pit, it appears there was more dumping of plant remains initially, which decreased over time. This is based on the very high densities in the middle two layers and the low density of carbonized seeds in the top layer of the pit. As noted before, there was a distinction in the disposal contents during the different dumping episodes. The first dumping episodes in the pit involved the disposal of high numbers of seeds and rachii of *Setaria* sp., seeds of *Linum* sp., and a range of weeds of which the SFL (Small Free Light) and SFH (Small Free Heavy) are dominant. The later dumping episodes involved the disposal of *Panicum* sp., *Setaria* sp., and importantly the discard of legumes.

Occupation III Harappan Pot, Mound

The contents of a Harappan pot discovered in Occupation III on the mound were floated for archaeobotanical materials. The carbonized seeds recovered include 79 caryopsis of *Setaria italica* and 2 *Digitaria* sp. seeds. A high percentage of the *Setaria* sp. seeds are fragments, but diagnostic. The pot was not sealed, and thus it is possible that the presence of the carbonized seeds in the pot contents could be a result of post depositional processes. For this purpose, the archaeobotanical contents of adjacent samples were examined. The archaeobotanical remains from the adjoining samples are comprised predominantly of *Setaria* sp. caryopsis, with *Eleusine* sp., *Setaria* sp. rachii, and *Trianthema* sp. occurring in low frequencies. The *Setaria* sp. seeds from these adjoining areas are also cracked but diagnostic. It is important to note, however, that the general deposits of the level in which the pot was located have relatively low densities (4 seeds per liter), while the pot contents have a significantly higher density of 81 carbonized seeds in 0.75 liter (101 seeds per liter). The significantly higher density in the pot is an indication that their presence is not due to post depositional processes. If the carbonized seeds infiltrated the pot through post depositional processes, then a similar density of carbonized seeds would be expected from the surrounding deposits.

Occupation II Refuse Ash Pit, Slope Unit

The refuse pit consists of loose ashy deposits with pockets of charcoal concentrations, burnt bone, and ceramic fragments. One side of the pit was dug into an adjacent stone wall of Occupation III. The nature of the deposits (presence of burnt bones etc.) suggests that the pit contents are domestic refuse and cleanings of areas related to possible cooking fires, as noted by the burnt bones. The pit was excavated by the author and a series of ashy dumpings (as opposed to one episode) were observed. The ash pit was excavated in several layers to facilitate stratigraphic differences. The densities of archaeobotanical remains from the lower layers of the pit are significantly higher than the upper layers. There is, however, no distinct pattern in the distribution of the various plant remains. The archaeobotanical assemblage of the pit is dominated by weed seeds (accounting for 60%) and millets (accounting for 35%). Legumes occur in low densities (5% of the assemblage) and are represented by *Vicia* sp. and *Lens* sp., although *Vicia* sp. occurs at a significantly higher frequency. The millet assemblage is predominantly represented by *Panicum miliare*. Among the weeds recovered from the pit, three weeds dominate the assemblage: *Urochloa* sp. (41%), *Digitaria* sp. (26%), and *Trianthema* sp. (21%). Two weed categories dominate the weed assemblage, those of SFH constituting 51% and the SFL category constituting 49%.

Occupation II Pits, Mound

Two pits excavated in units on the mound were sampled for their archaeobotanical remains. The contents of the two pits are distinct in that one pit is dominated by *Panicum miliare* and the other pit is dominated by *Setaria italica*. The pits are spatially distinct even though they are both in Occupation II. The sediment of both the pits was ashy dark gray brown with moderate amounts of charcoal. The pits are most probably related to domestic dumping and cleanings rather than dung fuel because of the high ratio of crop seeds to weed seeds. Miller (1984) argues that dung fuel contexts yield higher densities of weeds versus cultigens. The pit dominated by *Panicum miliare* (75%) has only two other archaeobotanical plant remains in the assemblage, that of *Cyperus* sp. (21%) and *Trianthema* sp. (4%). In contrast, the pit dominated by *Setaria italica* (48%) has a more heterogeneous assemblage with 5 other archaeobotanical components; *Setaria* sp. rachii fragments (2%), *Lens* sp. (2%), *Aizoon* sp. (2%), *Cyperus* sp. (6%), *Trianthema* sp. (2%), and unidentifiable seed fragments (40%). This patterned variation between the two pits is indicative of differential processing pertaining to the two crop plants, which will be addressed in the next chapter.

Occupation II Clay Feature, Mound

The last feature sampled for macrobotanical remains is an Occupation II clay feature excavated in one of the mound units. The clay feature is semi-spherical, set in the ground with slightly raised and distinct walls. The ashy sediments from this clay feature were floated for archaeobotanical remains. The sediments consisted of heavy clay clumps (most are probably the collapsed walls of the feature) and pottery fragments. The archaeobotanical assemblage is dominated by carbonized seeds of *Setaria italica* (60%). The other components of the assemblage occurring, albeit in low frequencies, are *Panicum miliare* (5%), a *Setaria* sp. rachis fragment (0.4%), *Lathyrus mungo* sp. (0.4%), *Lens* sp. (0.4%), *Echinochloa* sp. (0.4%), *Trianthema* sp. (2%), and unidentifiable seed fragments (30%). Although the nature of this feature is uncertain, it is probably a form of hearth (ethnographic and prehistoric examples of similar hearths occur in the Near East).

An important observation in the contextual patterning of the archaeobotanical materials of Babar Kot is that there is a distinct specificity of *Setaria italica* and *Panicum miliare* to different pit features. Only one millet dominates the contexts of a pit, and this indicates a specificity of activities pertaining to these two different millets. Therefore, only one of the two millets is being utilized dominantly in particular locales based on the evidence from pit features. A subsequent implication is that the two plants were treated as separate plant crops, albeit of varying degrees of importance. This has important implications for the reconstruction of the subsistence economy at the site, since more than one crop was being utilized as a food crop.

Summary of Babar Kot

The Babar Kot 1990-91 archaeobotanical investigations have revealed excellent archaeobotanical preservation and a high density of carbonized seeds. There are several conclusive statements that can be made about the Babar Kot archaeobotanical assemblage, most notably that there is a strong evidence of summer and winter crops. The summer crop of *Setaria italica* dominates the crop assemblage at Babar Kot, and this is similar to the site of Rojdi Phase C (Weber 1991). It is interesting to note these similarities between the two sites with respect to the role of millet cultivation in the subsistence economy of the region.

Summer cultivars are a very important and significant component of the macrobotanical assemblage of Babar Kot. These include 'millets' which dominate the assemblage throughout the occupation. In a majority of the cases, the millet crops *Eleusine* sp., *Panicum* sp., and *Setaria* sp. are not grown in the winter months in the present-day agricultural system of northwest India. On the other hand, the legumes *Lathyrus* and *Lens* are generally winter season crops. The legume seeds, however, occur in low densities and their use was probably limited. Therefore, it can be reasonably inferred that most of the subsistence cultivation was occurring in the summer monsoonal months, and limited cultivation in the winter months.

There are very distinct spatial differences with the slope units containing overall higher densities of carbonized seeds. There are spatial differences with respect to specific plants, with some only being recovered from the slope or only from the mound. The most significant spatial pattern is that the *Setaria* sp. spikelets with rachillas were recovered only from the slope units. This pattern is indicative of specialized activities related to food processing on the slope area and the prominent absence of such activities in the mound units.

Stratigraphically, later Occupation III has the highest densities of carbonized seeds. There is also an important association of carbonized seeds in significantly higher densities in feature contexts as opposed to general fill deposits throughout the occupation at the site. This association with feature contexts is particularly important because the seeds are indicative of specific human behavior and activities related to site maintenance and cleaning reflective of crop processing and food preparation. Determining whether the millet plants were used as fodder is a complicated issue, and other lines of evidence, such as carbon isotope analysis of the bone collagen (Reddy 1991a), are needed.

Summary of the Archaeobotanical Research

There is strong evidence for the use of summer crops at both Oriyo Timbo and Babar Kot. There are significant differences in the temporal, spatial, and contextual patterning of the frequencies of carbonized seeds recovered at the two sites. The archaeobotanical assemblage at Oriyo Timbo is dominated by weeds, although the millet seeds also occur in strong densities. Additionally the Oriyo Timbo archaeobotanical assemblage consists primarily of plant seeds, and has no rachii or any other plant parts. The Babar Kot 1990-91 archaeobotanical assemblage is indicative of both summer and winter crop usage, with the summer crop *Setaria italica* dominating the crop assemblage at Babar Kot. *Setaria* sp. rachis remains, which are strong indicators of crop processing stages, were also recovered from the site. In conclusion, archaeobotanical remains provide significant data to study past economy and subsistence systems. It is nonetheless important to realize the limitations of this data set, mainly in controlling for factors of deposition, preservation, and recovery. Study of the archaeobotanical data sets from Oriyo Timbo and Babar Kot has revealed that carbonized plant remains have several qualities that make them highly valued archaeological evidence. Once the limitations are realized and the attributes complemented by other relevant sources of data, such as ethnographic crop processing and farming studies as done in this study, then fresh issues can be explored.

6. If the Threshing Floor Could Talk: Testing the Ethnographic Models

In order to link the archaeobotanical remains to past human agricultural activities it is imperative to delineate and distinguish the predepositional and depositional processes (Jones 1987). The main focus of the study has been to discern the predepositional processes related to crop husbandry and crop processing sequences from harvest to use as food, fodder, or both. To discern the depositional processes, analytical and interpretive attention are focused on identifying and modeling the contexts in which preservation through charring of the plant remains (accidentally or intentionally) may be likely to occur, and the subsequent human behavior related to the discard of these materials. At the analytical level specific methods were applied to identify patterns in the ethnographic and archaeological data and evaluate their significance. In this chapter, interpretive theory (based on the ethnographic modeling) is used to assign behavioral meaning to the patterns observed in the macrobotanical assemblages, and it is also used to evaluate the relative merit of a series of possible interpretations.

Hypotheses and models of predepositional processes (specifically crop harvest and processing sequences), based on ethnographic observation, have been presented earlier. In this chapter, these hypotheses are tested against the archaeobotanical materials from Oriyo Timbo and Babar Kot. If patterns in the archaeological samples fall within the range of variation of the pattern of physical traces that resulted from a particular ethnographic activity (as summarized in the ethnographic models), one can then infer that a valid analogy exists. The degree of probability cannot be measured, but the strong correlation of the two patterns is indicative of a viable analogy. The correlation between the two patterns can be further strengthened by evaluating the merits of several other possible interpretations that may have created such a pattern in the archaeobotanical data. In other words, this process entails defining, limiting and accounting for other factors which could have caused related patterns in the archaeobotanical data. Such testing of multiple hypotheses in the application of ethnographic models to the archaeological data greatly strengthens inferences and reduces concerns about the weakness of direct analogy. In addition, since the archaeobotanical application of ethnographic models is based primarily on the physical characteristics of the distinguishable components (such as weeds, rachillas, and crop grain), it is not necessary that every species in an archaeological assemblage be present in an ethnographic study to make this comparison valid and informative.

Archaeological Application of Ethnographic Models

Application of the Models

The archaeobotanical assemblages from Oriyo Timbo and Babar Kot are significantly different in their compositions, nature, context, and recovery rates. The Oriyo Timbo assemblage consists primarily of summer crop seeds of *Eleusine* sp., *Setaria* spp., *Panicum miliare,* and *Phaseolus mungo*. In contrast, the Babar Kot assemblage is comprised of seeds, rachii, and spikelets of *Eleusine* sp., *Panicum* sp., *Setaria* spp., *Linum* sp., and legumes. The contextual, spatial, and stratigraphic patterning of the macrobotanical materials within each site is also significantly different. However, the questions posed for each assemblage are similar:

1. Were the crops cultivated at the site or nearby, or were they brought as grain from other locals?

2. If the crops were being cultivated, was it a primary subsistence activity, or were there other major subsistence activities such as herding cattle.

3. If cattle herding was also an important subsistence practice, what effect might seasonal migrations and pastoralism have on crop husbandry practices?

4. Were the millets used primarily for human

food, animal fodder, or both?

The investigations at the two sites focused on addressing the economic importance of millet cultivation, and the final objective of the paleoethnobotanical study is to present a working model for the economic importance of the millet crops *Setaria* spp., *Panicum* sp., and *Eleusine* sp. in the subsistence system at Oriyo Timbo and Babar Kot. The interpretation of the archaeobotanical assemblages as being more representative of a food-like or fodder-like assemblage, or both, is presented in the subsequent discussions.

To address whether crops were cultivated at each site by the occupants or brought in from elsewhere, the assemblages must have characteristics of particular crop processing stages that are indicative of these two distinct situations. For example, if crops were cultivated by the site occupants then it is highly probable that the lower processing stages of sieving and winnowing I occurred on site. In addition, if the cultivation was in the close vicinity, then one can expect (though it is not necessary) that some of the other lower processing stages (such as threshing and raking) also occurred at the site.

In contrast, if the occupants were bringing in the crop grain from elsewhere (i.e., they are not cultivating the crops but obtaining them through trade and exchange, or, they were growing them elsewhere and bringing them to this site in the form of highly processed grain, as a part of a seasonal round), then it is very unlikely that any lower processing stages would be observed at the site; instead, only the highest processing stages (such as pounding, winnowing II, and grinding) will be observed. This is because the crops are traded most often after their lower processing, particularly after the sieving stage but before the pounding stage.

It is necessary to first correlate the Oriyo Timbo and Babar Kot archaeobotanical assemblages to the lower or the higher processing stages and therefore, most comparable to a situation when crops are brought in from elsewhere or cultivated by the inhabitants. Then, the presence or absence of other possible factors (related and unrelated to crop processing) that could produce similar situations must be addressed. Once this question is answered, then the status of the millet crops (whether they were used primarily as human food or animal fodder) and the economic importance of millets for the Oriyo Timbo and Babar Kot inhabitants can be addressed.

The key characteristics of the lower processing stages and higher processing stages distinguished in the ethnographic study of Type B crops are distinct. Emphasis is placed here on Type B crops because the millets recovered at Oriyo Timbo and Babar Kot (*Eleusine coracana*, *Panicum miliare*, and *Setaria italica*) are all Type B crops whose harvests are most likely to resemble a Type II harvest. These Type B millets are processed similarly through all the stages from harvest to the final cleaning stages with the primary differences entailing how the grains are prepared for consumption. The lower processing stages include harvest (Type II method), threshing, raking, sieving, and winnowing I. Threshing and raking occur in the fields predominantly, and hence it is unlikely that their residues will be preserved or discovered archaeologically. Whether sieving and winnowing I occur more often at the home bases are dependent on several factors including the efficiency of threshing and raking, the state of the crop at harvest (dry or fresh), distance of the fields from the home bases, and the availability of labor. It is important to note that even if sieving was done at the fields, it is very often repeated at the home base immediately before the winnowing I process. Since these millets are small-grained seeds, winnowing by wind is not appropriate, and the processing is primarily winnowing by shaking for the separation of the chaff, straw and other larger light materials. The key characteristics of these lower processing stages of a Type B crop include straw, chaff, panicle heads, rachillas (of all kinds), spikelets, and grain with husk and weeds of all categories.

The higher processing stages of a Type B crop include the final winnowing I process, pounding, winnowing II by shaking, grinding, and food preparation. These stages are necessary in the processing of all Type B crops. These processes are in certainty done in the home bases, and therefore are more likely to be discovered archaeologically. As in the case of sieving, winnowing I is repeated immediately before pounding; however, the byproducts of the final winnowing I by shaking are predominantly comprised of undesirable crop grains and weeds (rachillas are processed out predominantly in the initial phases of winnowing I). The key characteristics of the higher processing stages of a Type B crop include crop grain with husk and without husk, and weeds of SHH (Small Headed Heavy), SFL (Small Free Light), and SFH (Small Free Heavy) categories. The crop products in general are cleaned of all the non-crop grain components (such as rachii, rachillas, and weed seeds) in the winnowing I by shaking stage. Grain without husk after the pounding stage includes a relatively high percent of cracked grains.

Storage of these Type B crops (*Eleusine coracana*, *Panicum miliare,* and *Setaria italica*) grain is widespread ethnographically. In the majority of cases fresh grain is seldom used as food, but has to be stored for some months before it is considered suitable for utilization. Consumption preference typically is given to grains that are older rather than fresh grain, mainly because of the enhanced cooking quality of stored grains. Customarily, storage and trade/exchange occurs after winnowing I and roasting (since the grain stores better with a decreased risk of fungal infections). This pattern is archaeologically significant since storage increases the probability of discovery, and roasting prior to storage often results in burning of some grains, which further increases the probability of survival. When crop grains are stored before utilization (as compared to immediate use) there is a higher chance that they will be exposed to accidental burning. Additionally, since the storage is typically in domestic areas (rather than in the fields which are harder to discover archaeologically) their discovery is further enhanced.

When Type B crop grain are most likely to enter the trade and exchange network can also be predicted and archaeological correlates identified. When the crop products enter the market is dependent on several factors including demand for the particular grain, labor involved to continue the processing, distance to market, and nature of the grain (whether it would hold up better in a husked or dehusked state). Ethnographic information gathered during this study indicates that for village communities grains of crops such as *Panicum miliare*, *Setaria italica* and *Eleusine coracana*, enter the market system after several different stages of processing. It is important to focus on modern village level practices to model for the Harappan subsistence systems, since they are more analogous than the larger 20th century market systems. Crop grains are often traded locally and non-locally immediately after winnowing I by shaking in the modern villages of Gujarat and Andhra Pradesh. At this stage the assemblages are devoid of all the larger non-crop grain components such as large rachii, panicles, straw, and large weeds but often still have some rachillas and small sized weeds. In some cases, marketing of crop grains is also done after the roasting, pounding and winnowing II stages, since roasting in the husked state aids the grain storage due to decreased exposure and likelihood to fungal attacks. At this point the grains have been dehusked and cleaned of all the chaff dust from the pounding stage, and the crop products are comprised only of crop seeds. Ground flour, the final product of the processing stages, is traded and exchanged only in rare cases.

The archaeobotanical crop seeds recovered from the two sites are direct indicators of specific losses that occurred at various stages of crop processing, mostly post harvest processing. This project's ethnographic study shows that crop grain losses can occur at harvest (through shattering of grains when a crop is harvested well after maturity - as seen in the 'Opportunistic' cultivation study and also at various post harvest processing stages). The magnitude of loss is primarily dependent on the particular processing stages, the method of processing, and the type of grain. Post harvest processing losses, particularly those that occur in the home bases, are of primary relevance to archaeologists since their preservation and discovery are high.

The analytical focus is on discerning different crop processing activities that existed in the past through the application of ethnographic models to archaeobotanical patterns. In this context explicit emphasis is placed on the distinguishing cultural transformation of the archaeobotanical record. Cultural transformations that affect the archaeobotanical record have been discussed eloquently by Miksicek (1987:224-230). Building on Schiffer (1976, 1983), and also Rathje and Schiffer (1982), he articulates archaeobotanical remains that are characteristic of de facto, primary, and secondary refuse. De facto refuse includes usable paleoethnobotanical materials abandoned in an activity locus such as seed caches; primary refuse entails trash discarded at the location of use, such as seeds recovered from cooking pits, and secondary refuse consists of trash deposited at some location other than the location of use, and this includes the vast majority of plant remains recovered from archaeological sites (Miksicek 1987). In the following section contextual patterns at the two sites will be discerned in terms of these cultural transformations.

Oriyo Timbo

The archaeobotanical assemblage from Oriyo Timbo consists of millet and legume crop seeds, and non-crop seeds (weeds), of which the non-crop/weed seeds occur in higher densities. Overall, among the millet crops, *Setaria* spp. and *Eleusine coracana* occur in higher frequencies. The weeds include the categories SFH (Small Free heavy), SFL (Small Free Light), and SHH (Small Headed Heavy) of which the SFH predominate (69%). The

Oriyo Timbo archaeobotanical assemblage consists exclusively of plant seeds, and has no rachii or other plant parts.

Local Cultivation versus Trade / Exchange

To address whether millet crops were being cultivated by the Oriyo Timbo inhabitants, the nature of the archaeobotanical assemblage, as being more indicative of lower or higher processing stages has to be determined (see glossary if clarification of these terms is needed). The presence of rachii would have been an indicator of lower levels of crop processing, and therefore of cultivation at the site. However, specific categories of weeds can be used as indicators of different stages of crop processing, and the categories present in the Oriyo Timbo weed assemblage (SFH, SFL, and SHH) shed light on what the archaeobotanical assemblage at the site is most indicative of, earlier processing stages or later processing stages. This information is used to determine whether the crops were cultivated at the site or brought in through exchange from elsewhere. In the ethnographic record, it was observed that crop products are traded and transported over relatively long distances most often as the products of the higher processing stages.

All three weed categories in the Oriyo Timbo assemblage are indicative of higher processing stages. Given the extant weed categories, (which are indicative of the higher processing stages) the lack of any components representative of the lower processing stages (such as panicles, rachillas, and large weeds), and the context of the recovery (in domestic areas), the Oriyo Timbo assemblage is interpreted as most indicative of higher processing stages such as winnowing II by shaking, most probably winnowing II product spillage or byproduct cleanings. Prior stages, sieving and winnowing I are also possible but since there are no large components the processing had to be of a finer nature.

Since the entire assemblage shows no indication of any lower processing stages, it is highly probable that these processes were not occurring at or near the site. At this point it is important to reiterate the close proximity of the seasonal stream (which would have been used in the cultivation of the crops) to the site. If crops were grown nearby, then the lower processing stages most probably would have taken place on site itself rather than away from the site in distant fields. This situation increases the probability that some components of these lower processing stages would have entered the archaeological record given the close proximity of the stream/cultivation area to the domestic area. Therefore, it is asserted that these crops were not grown nearby but that the crop grains were brought into the site from elsewhere. Since the water source (the seasonal stream) is located in the immediate vicinity of the site, it is unlikely that they would have conducted the lower processing stages anywhere other than within the site.

Another line of evidence that strengthens this argument of trading or exchange of the crops is the lack of contextual association between the crop seeds and the non-crop/weed seeds. For example, the distribution of millets and legumes reveals that Occupation II, specifically stratum 5A, has the highest density of crop seeds while non-crop/weeds occur in particularly low densities. The non-crop seeds and the crop seeds do not demonstrate any correlation of occurrence; that is, there is no pattern of associated occurrence of these two groups in the samples. This lack of co-occurrence is a strong indication that these two groups of plants were not deposited together. Specifically, a significantly higher percentage of non-crop (or weed) seeds occur in feature contexts related to food preparation, while crop plants (millets and legumes) occur in significantly higher percentages in non-feature contexts.

Additionally, the higher densities of non-crop plants in food preparation related contexts are indicative of their use as fuel. For example, in Trench S8 samples from contexts of ashy pit features (which are most probably cleanings from the food preparation areas and living floors) and scrapings above burnt hearth surfaces contain a high percentage of *Mollugo* sp. weed seeds.

Mollugo sp. seeds constitute the majority of the weed assemblage at Oriyo Timbo. One explanation for the presence of these seeds is that the *Mollugo* sp. seeds were contaminants in the food grains and were selected out and ultimately found their way into the discard/dumping contexts that are found archaeologically. However, there is no correlation between these seeds and the millet crop seeds. For example, Occupation II stratum 5A, which has the highest densities of millet and legume seeds, also has a low frequency for *Mollugo* sp. seeds. If the *Mollugo* sp. seeds were indeed contaminants/intrusives in the millet crop products, then one would predict a greater association in contexts and samples where the millet crop seeds were recovered. Therefore, a more probable explanation for the presence of *Mollugo* sp. seeds in these contexts is their use as fuel (intentional or unintentional), both as dried vegetative matter and as part of dung. Their inclusion into dung could be either through ingestion by the animal or as packing material.

It is important to reiterate that there is no specific relationship between the crop seeds and non-crop seeds, in that there is no association of occurrence of these two groups of seeds in any specific samples, that is there are no samples in which there is any indication of millet crops and weeds being associated together strongly. This evidence is very significant since it indicates that the subassemblages of crops and non-crops (or weeds) are not correlated to each other, and thus is another indication that they were not used or deposited together at the same time on the site as an assemblage. Therefore, the weeds did not enter the archaeological record of Oriyo Timbo as intrusives in the crop products and thus cannot be used as indicators of crop processing stages. This only reaffirms the interpretation that the crop grains were brought into Oriyo Timbo as a cleaned product, and there is no evidence of any significant cleaning of the crop products. This occurs only when the crop is at the highest of the processing stages, and is most indicative of having been brought to the locale in that state.

I proposed in the previous chapter that most of the crop seeds recovered at Oriyo Timbo are not from dung being used as fuel, but rather their inclusion in the archaeological record is evidence of their use as food. This interpretation is supported through two avenues of evidence. First, crop seeds occur in significantly lower densities in the features related to food preparation at Oriyo Timbo, while non-crop/weed seeds occur in significantly higher densities in these contexts. Therefore, there is an apparent association of the non-crop/weed seeds with the food preparation related features, while the crop seeds show a significant association with non-feature contexts. If indeed the crop seeds recovered from the sediments of Oriyo Timbo were dung-related, then it is predicted that they would show a higher density in the food preparation related features where dung would have been used as fuel. Plastering of floors was not observed at the site, and this eliminates the inclusion of the seeds from dung-plastered floors. The second line of evidence that supports the interpretation of the crop seeds not being dung-related, is that if the crop grains were acquired by the Oriyo Timbo inhabitants through trade or exchange or grown elsewhere (outside the locales) and brought into the site as highly processed assemblages for human food, then it is highly unlikely that they would have used grain to feed animals. This would have been an extremely costly strategy when wild forage would have been available in the surrounding vicinities of the site.

Even though the archaeological patterns fit in well to the expectations of ethnographic models, it is very important that other factors (related and unrelated to crop processing), which could produce the patterns observed in the Oriyo Timbo archaeobotanical data, be addressed. Two additional situations, which could have produced patterns indicative of higher processing stages, are addressed below.

One factor that could skew the interpretation that the Oriyo Timbo inhabitants only conducted higher processing stages is that the lower processing stages occurred outside the site, and, therefore, cannot be identified. Although this is conceivable there are several reasons for considering that the probability is very low. First, even if the initial processing stages (threshings) occurred outside the site, one would still expect that some of the subsequent stages (such as sieving and winnowing I) occurred within the site. None was discerned. Second, even if the occupants were cultivating, it was undoubtedly done in close proximity since the streambed is within meters of the site. Given this situation it is highly unlikely that the lower processing stages would occur away from the site.

Another factor which could skew the interpretation that the Oriyo Timbo archaeobotanical assemblage is representative of higher processing stages (winnowing II product spillage or byproduct cleanings most probably) is the absence of rachii and rachillas due to them not surviving into the archaeological record and inhibiting the identification of the lower processing stages. Experimental studies on the charring and endurance of rachii, rachillas, and spikelets of the various millets showed that these components endure burning and there is a high possibility of their survival into the archaeological record. Even if one assumes they did not survive, in the specific case of Oriyo Timbo the weed categories can still be used as isolating variables for lower or higher crop processing stages. It is highly improbable that a situation similar to 'Opportunistic' Cultivation (where the crop fields are sterile for weeds) would be occurring at Oriyo Timbo since the area lacks the appropriate geological context that requires a heavy clay flood plain.

Thus, in summary, the archaeobotanical assemblage at Oriyo Timbo is dominated by weeds, with millet seeds also occur in strong densities. These weeds are critical in defining the nature of the assemblage with respect to local or non-local cultivation. Additionally the Oriyo Timbo archaeobotanical assemblage consists primarily of plant seeds, and has no rachii or any other plant parts. Their presence would have been an indicator of lower levels of crop processing, and therefore of cultivation by the site occupants. Given the nature of the assemblage and the lack of storage heaps or facilities, and the high proportions of weeds to summer crops, it is suggested that the summer crops were not cultivated by the site occupants. The crop seed archaeobotanical assemblage of Oriyo Timbo is most indicative of a situation where the crops were being brought into the site from outside, as essentially cleaned (of weeds, rachii, rachillas, etc.) products. There are no other factors (related and unrelated to crop processing) that could produce the patterns observed in the Oriyo Timbo archaeobotanical data. The crop grains could have been brought into the site by the occupants, from their settlements in other parts of the general area, or obtained through local exchange.

Economic Use of Plants: Food or Fodder

To interpret whether the Oriyo Timbo archaeobotanical assemblage shows characteristics of being a food-like assemblage, a fodder-like assemblage or a combination of both, a series of issues must be addressed. The characteristics of food-like and fodder-like assemblages need to be elucidated and summarized based on ethnographic research. Then the Oriyo Timbo material can be correlated to the two different or to a combination of the two assemblages. The entire archaeobotanical assemblage and the sub-assemblages of crop seeds and the non-crop/weed seeds (particularly since they do not show any association and correlation of occurrence) from the site need to be considered.

Food and fodder assemblages are quantitatively and qualitatively distinct. A food-like assemblage is one that is closely related to being used as human food (immediately or through later processing). In the ethnographic models all products of the crop processing stages, especially in the higher processing stages, are found with food-like assemblages. Typically they include dominant proportions of crop grains (husked or dehusked), along with weed categories whose morphological characteristics are similar to the crop seeds (SFH and SFL). They also include significantly smaller sized components than the lower processing stages. The higher stage assemblages are rarely fed to animals as fodder since they are the products that have already been extensively processed for ultimate use as human food. In the rare ethnographic cases in India, when animals are fed these products, it is for very specific reasons and of very short duration. For example, feeding of such products during specific festivals, and to pregnant cows particularly if they are malnourished and need improved diets.

A fodder-like assemblage is not always derived from crop processing stages, since it includes the byproducts of the different stages and also wild forage collected intentionally for animal feed. Only the former situation can be determined with certainty in the archaeobotanical data, since the latter assemblage can be often confused with fuel. However, bone chemistry can be used to gain further insight into the diet of the animals. A fodder-like assemblage related to crop processing may include any or all byproducts from the different processing stages. As indicated in the models, the ethnographic crop byproducts from all stages are fodder-like assemblages. They typically include significantly lower densities of crop grains, and are dominated by the non-grain components of the crop (such as straw, panicles, chaff, rachii, rachillas, undesirable spikelets, and weeds).

Based on the patterns observed in the Oriyo Timbo archaeobotanical assemblage, the overall assemblage is interpreted as consisting of food-like assemblages, and fodder-like assemblages that are not derived from crop processing products or byproducts. Specifically, spatial contexts of the samples with millet and legume crop seeds are distinctly correlated with spillage and cleaning episodes. As discussed earlier, they are comprised of only seeds with no rachii, rachillas, and spikelets. In addition, there is the distinct absence of strong correlation between the crop seeds and weed seeds, indicating they were not deposited together. The millet assemblage is thus interpreted as being derived from a relatively high processing stage, most likely a crop product, and thus very much a food-like assemblage.

The non-crop/weed assemblage however, is more similar to either a non-crop processing fodder-like assemblage, or fuel for food preparation assemblage. It is not related to byproducts or cleanings from the food-like assemblages. This assertion is supported by the lack of strong association of weed seeds to crop grains and non-grain components such as rachii, rachillas, and spikelets. Additionally, the non-crop/weed seeds were found to be significantly associated with the food preparation features. This could be attributed to several processes, including crop product cleanings swept into hearths, their incorporation into dung that was then used as fuel, and use as fuel. As established earlier, the non-crop/weed seeds are not considered a crop processing byproduct assemblage. To reiterate, there is loss of crop grain (of the smaller immature crop grain) during any crop product cleaning, and such grains would be associated with contexts where byproduct cleanings of non-crop/weed seeds occur. The absence of this association affirms that such byproduct cleaning did not occur at the site.

In summary, given the lack of evidence for lower crop processing stages, it is argued that cultivation of millet crops was absent at Oriyo Timbo. It is, however, possible that if the Oriyo Timbo inhabitants were seasonal occupants at the site, they could have cultivated the millets elsewhere in another region, and brought them to the site at the time of occupation as highly processed crop products at the higher processing stage levels. Nonetheless, by demonstrating the absence of millet cultivation at Oriyo Timbo one of the major objectives of this study has been achieved.

Model for Plant Usage and Subsistence at Oriyo Timbo

Based on these results, the subsistence system of Oriyo Timbo is interpreted as primarily geared towards animal husbandry, and millet cultivation was absent. However, an integral aspect of the subsistence economy was the use of millets as human food, despite these grains being traded items or being brought into the site from elsewhere. The location of the site optimized the proximity to a water supply (which is crucial when herding a large number of animals), close distance to contemporaneous Late Harappan farming villages, and the readily available wild forage in the area. Plant utilization at Oriyo Timbo was extensive as illustrated by the plant usage model (Figure 6-1). The Oriyo Timbo plant usage model presents the plants that were intentionally brought into the site, and they are comprised of crop plants, wild forage, collected wild food plants, fuel, and building materials. The human subsistence system most likely included the use of millet grain and legumes, supplemented by the collection of wild plants such as *Chenopodium* sp., *Amaranthus* sp., and *Dactyloctenium* sp. Animal meat consumption (both domestic and wild) contributed significantly to the diet of the inhabitants (Rissman and Chitalwala 1990). Given the importance of animal husbandry at Oriyo Timbo, wild forage collection for animal feed was probably an important subsistence related activity. Some wild forage collected might have included *Mollugo* sp., *Trianthema* sp., *Phyllanthus* sp., *Cyperus* sp., *Digera* sp., *Digitaria* sp., and *Urochloa* sp. as represented in the archaeobotanical assemblage of the site. *Carex* sp., and *Solanum* sp. are not edible and were probably used as fuel.

If Rissman's (1985) seasonality data are accepted, then Oriyo Timbo was occupied March through July, which does not coincide with millet harvest seasons (which are most often during the months of October and November). Therefore, it is unlikely that the occupants could have obtained millet crop processing byproducts from neighboring villagers for use as fodder, unless the byproduct was stored by the farmers for later use. In any case, the consumption of millet crop processing byproducts as fodder was most probably immediate and hence cannot be identified through archaeobotanical research.

The plant usage model shows millets being brought into the site (obtained through trade and exchange) at winnowing I or II stage (Figure 6-1). The usage and disposal of the plants (both crops and wild plants) are complex, but several processes have been elucidated, including roasting, winnowing after roasting and winnowing II. The disposal of the crop seeds as primary and secondary refuse has been identified and includes hearth contents, and trash/dump pits. Although the model presents several processes, all the processes need not have occurred in one locale or throughout the duration of site occupation. The presence of these processes during the Late Harappan occupation of Oriyo Timbo, however, reiterates that although millets were the primary food crops, human diet was also supplemented by wild plant resources.

Further insight into the character of pastoralism and animal husbandry at Oriyo Timbo can be gained by drawing on ethnographic observations of modern pastoralists in Gujarat (specifically their interaction with farmers, and fodder acquisition). Traditional agriculture is an important subsistence economy in this region of India, but it is usually complemented by two different forms of pastoralism: pastoral nomadism and semi-nomadic pastoralists. The definitions of these two forms of pastoralism have been developed by Khazanov (1984). His classification is an economically ori-

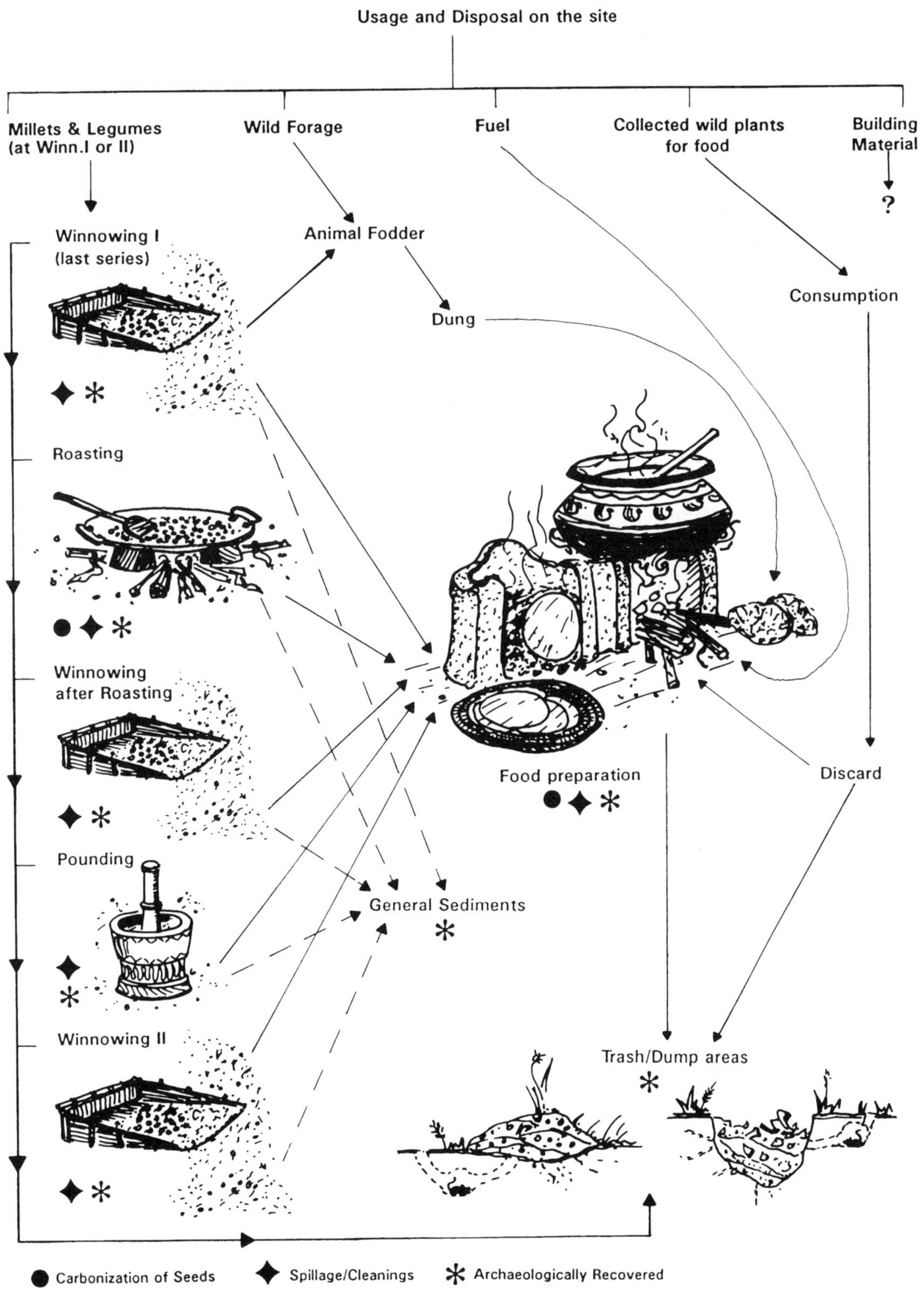

Fig. 6-1. Oriyo Timbo plant usage model.

ented system focusing on the degree of pastoral dependence on agriculture. Pastoral nomadism is characterized by the absence of agriculture altogether, with grains and other domestic crops obtained from other groups. In contrast, semi-nomadic pastoralists are characterized by pastoralism as the major economic activity; however, agriculture occurs as a secondary and supplementary activity. In Gujarat both these types of pastoral groups are observed, with two main types of pastoralists within each group: cattle and goat herders, and camel herders. Seasonality is an inherent ecological factor in the region, and the seasonal flow of the rivers and other ecological constraints in the area, such as the seasonal availability of wild grasses, contribute to making pastoralism an important subsistence economy.

The pastoral nomads rarely practice agriculture, even on a seasonally limited basis, and survive through animal husbandry. Pastoral communities making their living on animal breeding are a distinctive feature of rural Gujarat. Being mostly a landless group, their stock is dependent on community grasslands for survival. Therefore, migrations are determined primarily by pasture availability. The main annual movements are from the north and northwest towards the more fertile and forested area to the east and southeast in the post monsoon months (Patel 1977). The seasonality of this movement has a direct effect on fodder acquisition and its utilization by stock. Often the pastoralists migrate with the entire family and stay in specifically chosen areas for nearly six months. During their migration, the duration of the temporary camps is usually dependent on how long local inhabitants allow them to stay at one place. En route they often rent plots of fields from farmers, camping directly on these field plots and paying their rent with animal manure.

In some cases, semi-nomadic and semi-sedentary pastoralism is also observed in Gujarat, with permanent pastoral settlements situated within villages, and these pastoralists occasionally own land to cultivate for subsistence. These semi-nomadic and semi-sedentary pastoralists practice agriculture, both for human consumption and animal fodder, and the extent of its importance varies depending on the group. In both cases, the pastoralists still participate in seasonal migrations. Very often part of the family (the elderly and often also women) remains in the village with some of the herd, usually the goats, while the rest of the family takes the cattle southward for pasture grazing in the summer months (March through June).

Given the juxtaposition of larger sites with more substantial remains (villages?), and smaller more ephemeral sites (pastoralist communities?) in the area during the Late Harappan Phase in Gujarat, it is conceivable that the occupants of Oriyo Timbo may have participated in a related interaction system. The site of Oriyo Timbo could have been a part of a seasonal round for a pastoral-nomadic community who engaged in trade and exchange with the local agriculturalists for food grains. Or, Oriyo Timbo could have been the seasonal camp of semi-nomadic or semi-sedentary pastoralists engaged in a mixed and integrated economy of farming and pastoralism, and thus bringing with them highly processed crop products on their seasonal migrations to the site. The existing regional data for the Gujarat Mature and Late Harappan traditions, however, are not detailed enough to test such hypotheses at this time. Nonetheless, it is important to be cognizant of such economies and settlement patterns while suggesting alternative explanatory models for the occupation of Oriyo Timbo.

Fodder acquisition and utilization among the pastoralists in Gujarat are very specific and highly organized. There are distinct differences in fodder usage for cattle, water buffalo, and goats. In general, cattle and water buffalo are fed both dry and green fodder, while goats and sheep are fed only green fodder. *Pennisetum typhoides* and *Sorghum bicolor* byproducts obtained dry from farmers usually in exchange for animals products (such as milk, butter, fur or manure) are fed to the cattle, but not to sheep and goats. Additionally, cattle are fed *Pennisetum typhoides* chaff and straw, while water buffalo are given only *Sorghum bicolor* straw and chaff. The reason for this pattern is not entirely clear. There is a significant difference in the crude fiber and crude protein between the straws and chaff of the two millets, and this may be the determining factor in the effectiveness of different fodders for cattle versus buffalo.

Goats and sheep are not fed dry fodder since it negatively affects milk production, and hence green fresh fodder is always preferred. This is a surprising and important observation, since goats are capable of existing on a nutritively poorer diet, but the pastoralists of Gujarat never feed their goats dry fodder. Often the green parts and the seeds of field weeds such as *Trianthema portulacastrum*, which often grow in millet crop fields, are selectively collected and fed to goats (and cattle). Some common wild fodders, such as *Eremopogon foveolatus* and *Apluda aristata*, are collected and fed to cattle and goats.

Fodder utilization by pastoralists is very seasonal. For example, selective collection of the wild fodder *Cressa cretica* (popularly called Bokhna) is common in parts of North Gujarat. *Cressa cretica* is readily available in November/December since it needs a lot of dew for its growth. It is used only as a green fodder and it is the first preference when it available since it enhances milk production. As it cannot be stored as dry fodder, it is used profusely while available, after which other fodders are used. In the months when this wild fodder is not available, other wild fodders and millet byproducts such as *Pennisetum typhoides* straw and chaff are used. *Brassica* sp. and *Amaranthus* sp. are two other plants (cultivated and wild varieties) used and selected specifically by pastoralists for animal fodder during other portions of the year. Such seasonality in fodder intake is common in northern Gujarat. Even in other areas of Gujarat, the pastoralists follow similar patterns of fodder utilization, feeding the selectively collected wild fodder during the rainy season of July through August. Millet byproducts such as *Pennisetum typhoides* straw and chaff are used in the summer months of March through June. Moreover, they often have to buy *Pennisetum typhoides* straw when other fodder is not available.

My ethnographic research has noted interaction between pastoralists and settled farmers in different parts of Gujarat, and this interaction involves considerable exchange of food and fodder between these two complementary groups. Such a pattern and exchange system have important implications for Oriyo Timbo, and it is possible that such interaction might have occurred between the inhabitants of Oriyo Timbo and settled farming neighbors. In ethnographic situations, pastoralists often exchange animals and animal products such as milk, butter, and processed butter for crop grains (for human consumption) and crop byproducts (for animal fodder). Crop byproducts are fed to the cattle (but not to the goats) when wild forage is scarce. In general, there is specific selection of wild forage with attention being focused on the inherent 'milk-producing' qualities of the wild plants. This is important because milk production defines the success of a pastoralist economy. For example, green fresh forage is preferred over dry plants, since it enhances lactation, while dry forage has a tendency of limiting milk production in some animals.

Based on the previous discussion, four main ethnographic observations regarding Gujarat are relevant to interpreting the occupation of Oriyo Timbo and in general for understanding the Late Harappan in the area:

1. There is an interdependence (complementarity) between the agriculturalists and the pastoralists.
2. There is considerable seasonality in fodder acquisition and utilization by both pastoralists and farmers. The emphasis on fodder is dependent on how important the animals are for the economy (apart from being load animals).
3. There is selective fodder acquisition by the pastoralists, mainly dependent on the effect the fodder has on milk production.
4. In some cases, the pastoralists are involved both in animal husbandry and small-scale cultivation. The emphasis varies depending on the year and specific needs.

The interplay of farming and herding within an uncertain semi-arid environment is characteristic of Gujarat. Rissman (1985:366) proposed that a lack of security in regular yields prehistorically would have led to the build-up of cattle herds to ensure a successful subsistence economy. Moreover, farming practices might have been temporarily abandoned and a period of specialized pastoralism developed in the face of continued agricultural uncertainty. Conversely, increasing confidence in cultivation would allow a household to dispose of animals to a level at which some degree of village life would again be viable. Observations of contemporary pastoralists and agriculturalists of Gujarat, however, do not support this scenario. Instead, a whole range of possible adaptations, from pastoral nomads with no cultivation to semi-nomadic pastoralists who practice some cultivation to settled farmers is documented. Therefore, if cultivation failed prehistorically (as suggested by Rissman) the succeeding economy need not necessarily have been pastoralism to the total abandonment of cultivation. Instead, it is plausible that a combination of the two economies took place. Additionally, there are no archaeological data (climatic or archaeobotanical) that indicate major failure in the agricultural systems of the Gujarat Late Harappan.

Thus, the ethnographic evidence demonstrates that pastoralism and agriculture can coexist and need not be successive and alternate economic strategies. I propose two alternate possibilities for Gujarat Late Harappan settlements: pastoral nomads and agriculturalists, and semi-nomadic pastoralists conducting small scale agriculture or semi-sedentary pastoralists supplementing their dominant subsistence economy of agriculture with pastoral activities. The first case occurs when there is a constant interplay between pastoralists and agriculturalists in terms of balancing resources. The second case is when there is a complex interlacing of the two economies, pastoralism and agriculture in the same group. In the context of Oriyo Timbo, I suggest that it is probable that any of these interactions and integration of economies were present.

At Oriyo Timbo, animal husbandry appears to have been the key subsistence practice. The occupants were thus obliged to concentrate their subsistence related activities on fodder acquisition, either wild forage collection, and also they may have exchanged with the neighboring farmers for crop grains (for human food) and crop byproducts (for animal fodder). Rissman (1985, 1985b, 1987) and Ryan (1990) both suggest that the age profiles at death for cattle show predominantly young adults and juveniles, with young calves in small numbers. They suggest that this is a possible reflection that this was a fallow herd with few lactating females. The slaughter of young juveniles and young calves implies that animal meat was an important component of the diet and possibly used in exchange with farmers.

In summary, when the archaeobotanical and archaeofaunal studies are considered together, it is most probable that Oriyo Timbo inhabitants were part of a larger, more complex regional settlement and subsistence system. The inhabitants of Oriyo Timbo were not self-sufficient. Instead a complementary subsistence system existed between the pastoral nomadic, semi-nomadic or semi-sedentary pastoralist inhabitants of Oriyo Timbo and the neighboring farming villages, based on exchange of human food in both directions and the acquisition of fodder by the pastoralists.

Babar Kot

To address the economic importance of millet cultivation at the site of Babar Kot, it is imperative to first determine whether the millet crops were being cultivated by the site occupants. Subsequently, determination of the role of these crops (as food for humans, fodder for animals or both) in the subsistence economy, and the degree of reliance on them are necessary to elucidate their economic importance to the occupants of Babar Kot. Then a working model for the plant usage and subsistence at the site will be presented that summarizes the findings of this research.

Local Cultivation versus Trade / Exchange

The first step in examining the issue of whether crops are brought in from elsewhere or cultivated by the inhabitants is to correlate the archaeobotanical assemblage of Babar Kot to the lower or higher processing stages. First, the Babar Kot archaeobotanical assemblage is correlated to a situation of cultivation by the inhabitants versus crops being brought in from elsewhere, then the presence or absence of other factors which could produce similar patterns in the archaeological record need to be addressed.

The archaeobotanical assemblage from the site consists of millet seeds, legume seeds, oilseeds, millet spikelets with rachillas and rachis fragments, and the four weed seed categories. Overall, crop seeds dominate the assemblage. The presence of spikelets with rachillas and rachis fragments, however, is crucial in discerning the nature of the assemblage. These two components are not representative of most of the higher processing stages such as roasting, pounding, winnowing II, or grinding. As shown in the study of crop processing stage products and byproducts of *Panicum miliare* and *Setaria tomentosa*, rachis fragments are indicative of lower processing stages. Rachis fragments are processed out predominantly through sieving. Spikelets with rachillas are also indicative of lower processing stages, although they occasionally occur in the final winnowing I process of the higher processing stages. They are predominantly processed out through the sieving and winnowing I process. Presence of these two components is a strong indication of two situations: that these two processes were conducted at the site, or that these processes had yet to be conducted. In order to determine which of these two situations is most probable, context and association must be examined, for example; spikelets and rachii occur in significantly higher frequencies in feature contexts and in association with crop and weed seeds.

In addition to the rachillas and the rachii, weed categories are also used to identify the crop processing stages represented. The Babar Kot weed assemblage includes four categories: those of SFL, SFH, SHH, and BFH. The ethnographic modeling presented in chapter 4 has demonstrated that specific categories of weeds can be used as indicators of different stages of crop processing, and the categories present in the Babar Kot weed assemblage (SFL, SFH, SHH, and BFH) can be used to shed light on whether the archaeobotanical assemblage is most indicative of an earlier or later processing stage. In general at Babar Kot, weeds occur in significantly higher densities in feature contexts. Thus, their presence is correlated to specific human behavior and activities, most probably linked to crop processing and food preparation. The SFL category of weeds dominate the assemblage accounting for 69%, with SFH weeds well represented (30%). The remaining two categories (SHH and BFH) together occur in very low percentages (0.6%). They cannot be ignored, however, without comment since they occur exclusively in feature contexts at Babar Kot and thus are a result of human activity. The BFH category is represented by wild *Ziziphus* sp., which are not necessarily weeds since they are commonly collected for human consumption. Its presence in the archaeobotanical assemblage at Babar Kot is attributed either to intentional collection and use as human food unconnected to crop processing, or possibly weeds being processed out from crop products. The SHH weed category is represented exclusively by *Phyllanthus* sp. which is a common weed in millet fields and is also fed as green cattle fodder.

The four weed categories (including the BFH category) that define the Babar Kot weed assemblage are indicative of both lower and higher processing stages. It is particularly significant that the SFL weed category (represented by *Cyperus* sp., *Eragrostis* sp., *Phalaris* sp., *Polygonum* sp., and *Urochloa* sp.) occurs in the highest frequency. These weeds are morphologically distinct from the crop seeds of *Panicum miliare* and *Setaria italica*, which are essentially SFH seeds. Ethnographic modeling has demonstrated that weeds which have similar characteristics to the crop grains are processed out in the higher processing stages while the weeds which are most distinct from the crop grains are processed out in the lower processing stages. Applying this to the Babar Kot weed assemblage, the higher ratios of the SFL weeds are a positive indicator of the lower processing stages.

Complementing this on-site processing is the strong presence of the SFH (and a lesser presence of SHH) category, which is an indicator of higher processing stages. However to distinguish the two levels of processing more securely, the weed category data have to be used in conjunction with the data on rachis fragments and spikelets with rachillas. Given the weed categories, the presence of rachis fragments and spikelets with rachillas (the key components of lower processing stages), and the contexts and features of archaeobotanical recovery, the Babar Kot assemblage is interpreted as indicative of both lower and higher processing stages. These include the stages of sieving, winnowing I, winnowing II, and possibly pounding (due to the presence of cracked millet grains in specific contexts). Overall, the assemblage demonstrates that a considerable amount of crop cleaning took place at the site and this sheds light on the nature of distinct and important activities related to food preparation and consumption.

Given the identification of lower and higher crop processing stages in the archaeobotanical assemblage of the site of Babar Kot, it is now appropriate to comment on the implications of this conclusion. The presence of lower crop processing stages is indicative of a certain degree of proximity to the first stage of crop processing, harvesting and threshing, which occur only if there is cultivation at the same locale. If the crops were obtained through trade or exchange, the lower processing stages would not be observed. This leads to the ultimate deduction that the crop grains are being cultivated in the general environs of the site, most probably by the site occupants themselves. The presence of the higher processing stages indicates that the cultivated crops were undoubtedly used as human food. The possible use of the crops as animal fodder, however, needs to be explored further.

The original source of the crop seeds in the samples needs to be established for the Babar Kot assemblage; particularly whether they are result of human food discard and accidents, or incorporation into dung that was used as fuel. This is particularly important since there was a strong association between high crop seeds densities and feature contexts. Two avenues of evidence support the interpretation that the Babar Kot crop seeds are for use as human food and not present due to incorporation into dung that was subsequently used as fuel. First, only one of the features at Babar Kot is a context where dung would be used as fuel, a hearth/cooking feature. All other features are dumps or trash pits related to domestic activities and could be the result of cleaning hearths and cooking features. The crop seeds recovered from these trash dump contexts are very unlikely to be dung-related, since the burning of the dung as fuel would completely destroy them. Instead, they are interpreted as crop product cleanings associated with food preparation. Therefore, given the nature of the feature contexts the crop seeds are interpreted as not related to incorporation into dung, instead their inclusion is indicative of their use as human food.

A second line of evidence that supports this interpretation includes the strong evidence of lower and higher crop processing activities at the site. This further supports the assertion since these activities are not performed on crops cultivated only for animal fodder. The majority of situations where a crop is cultivated as animal fodder it is harvested before seeding and then fed as green fodder. More often crops are cultivated to use the grain as human food and the byproducts are then fed to animals as fodder. Therefore, given both of these lines of evidence it can be asserted with certainty that the crop seeds were not incorporated into the Babar Kot archaeobotanical assemblage through the use of dung as fuel.

In addition to the identification of key elements of different levels of crop processing stages, several spatial and contextual patterns distinguished from the Babar Kot data strengthen the argument that these crop-processing activities are present. As mentioned earlier, there is a strong association between weeds and millet seeds. In general, both plant groups (crops and weeds) occur in significantly higher densities in feature contexts. This directly connects these remains to human activities related to crop processing and food preparation. There is a distinct spatial pattern in the distribution of *Setaria* spikelets with rachillas. They occur exclusively in the feature contexts of the slope units at Babar Kot. In addition, *Setaria* rachii occur in significantly higher frequencies in features within the slope (and to some extent on the mound as well). This pattern reveals that lower processing activities related to the cleaning of *Setaria* crop products were occurring predominantly in the slope area. Thus, it appears that this area of the site was where these processing activities were concentrated. *Linum* sp. seeds also have the same spatial and contextual association. *Linum* sp. seeds, all showing rupture signs on the narrow ends, occur exclusively in the slope units and in significantly higher densities in feature contexts. Although speculative, it is plausible that this pattern is indicative of a processing activity either for extracting oil or for cleaning the seeds for human consumption.

Another spatial distinction related to crop processing activities entails the distribution of the various weed categories. The mound weed assemblage is characterized by the dominance of two categories: SFL (49%) and SFH (50%). The slope weed assemblage reveals a dominance of SFL (66%) weeds with the SFH (32%) weeds occurring in lower but moderate densities. In addition, the slope weed assemblage also includes a significantly higher frequency of SHH and BFH weed categories (1.3% versus 0.4%). Ethnographic models indicate that the crop product and byproduct weed assemblages which are dominated by both SFL and SFH weed categories (as seen in the mound weed assemblage) are primarily those related to lower processing stages such as sieving and winnowing I. However, in contrast the later series of winnowing I are dominated by the SFL weed category (as seen in the slope weed assemblage).

The combination of weed and crop spatial distribution reveals a distinct pattern related to crop usage and processing at Babar Kot. The millet crop seed distribution in the mound area of the site includes *Setaria italica* (78%), *Panicum miliare* (20%) and *Eleusine coracana* (1%), while the slope area has *Setaria italica* (94%) and *Panicum miliare* (6%), but no *Eleusine coracana*. The significantly higher percentage of *Setaria* sp. in the slope implies greater usage (and processing) in this area. In addition, the exclusive presence of *Setaria* spikelets with rachillas, the significantly higher frequencies of *Setaria* rachii, and associated weed categories in the slope indicate there is a higher occurrence of both lower and higher processing activities related to *Setaria italica* in this part of the site.

On the mound, the presence of SFL and SFH weed categories indicative of lower processing stages associated with *Panicum miliare* indicates that lower processing stages of *Panicum miliare* took place on site. This implies local cultivation by the inhabitants. This argument is strengthened by the association of *Panicum miliare* seeds and both the SFL and SFH weed categories specifically with feature contexts. Thus two important inferences are made. First, both *Setaria italica* and *Panicum miliare* were cultivated near the site and processed on site. Second, there is spatial distinction in where the processing of the two crops occurred. *Setaria italica* was predominantly processed on the slope while *Panicum miliare* was more often processed on the mound.

A series of temporal patterns are also important to consider in examining the issue of local cultivation and processing. There is a significant difference in the carbonized seed densities from the three occupations at the site. Occupation III has a significantly higher density of carbonized seeds. This pattern could indicate either differential usage of plants through time, or temporal variations in crop processing. At Babar Kot, it is argued that the differential usage of plants through time is most probable. First, all the plants recovered at

Babar Kot (regardless of occupational level) are Type B crops which all require similar types of processing. Both of the major crops, *Setaria italica* and *Panicum miliare*, are hulled crops in that their grains are enclosed by tough glumes and the release of the grain requires more than threshing. Therefore, similar crop processing is required for both the crops.

Second, even though Occupation I has only *Panicum miliare* and there are no rachii or spikelets with rachillas, the weed assemblage from this occupation is comprised of equal percentages of SFH and SFL categories. These are indicative of lower processing stages. Thus, it appears that in Occupation I the site inhabitants were only (or predominantly) cultivating and processing *Panicum miliare*. In Occupation II, however, *Setaria italica* was introduced and its cultivation is indicated by the presence of rachii and spikelets with rachillas (both indicative of lower processing stages). A noted decrease of *Panicum miliare* occurs in Occupation II. In Occupation III, there is steady and significant increase in *Setaria italica* seeds, rachii, and spikelets with rachillas. *Panicum miliare* increases in density during Occupation III but still at significantly lower density than *Setaria italica*. The *Panicum miliare* seeds recovered from Occupation II were recovered in significantly higher densities from feature contexts indicating direct association with food preparation and consumption. So even though there is an obvious decrease in the use of this crop in Occupation II, its fundamental usage as human food continues. This decreased use of *Panicum miliare* in Occupation II is most likely related to the addition of *Setaria italica* and not due to variations in processing pathways.

Setaria italica occurs in significantly higher densities in Occupation III. There is distinct increase in the usage of this crop at the site from Occupation II to Occupation III, as seen by the increase in seeds, spikelets with rachillas, and rachii of this crop from various contexts. The increase is reflective of an increase of usage and it does not imply different processing activities, since there are very limited ways to process a Type B crop. Thus, there is distinct variation in plant usage over the occupation span of Babar Kot.

In addition to the variation in plant usage over the temporal span of the site, there is a distinct pattern in the stratigraphic distribution of the weed categories. The weed assemblage of Occupation I is almost equally represented by SFH (47%) and SFL (53%) weeds, while Occupation II and III weed assemblages have SFL category in dominating frequencies (71% and 73% respectively). Ethnographic modeling indicates that crop product and byproduct weed assemblages represented by both SFL and SFH weed categories (as seen in Occupation I) are primarily indicative of lower processing stages (such as sieving and winnowing I). In contrast, the later series of winnowing I have the SFL weed category in dominating frequencies (as seen in Occupation II and III). Coupling this temporal patterning in distribution with the significant presence of spikelets with rachillas and rachii in Occupation II and III, confirms the presence of activities related to several lower processing stages in these latter occupations.

Of all the crops utilized by the Babar Kot the use of *Eleusine coracana* is the most problematic. Only eight caryopsis of this crop plant were recovered from the site and all occur in the mound from non-feature contexts of Occupation II and III. Recovery of *Eleusine coracana* seeds/caryopsis was significantly low from Babar Kot (as compared to Oriyo Timbo), constituting less than 0.5% of the millet assemblage (while the Oriyo Timbo *Eleusine* sp. constituted 15% of the millet assemblage). The low densities of the plant reflect its lower usage and also possible usage as green fodder. Two issues must be addressed to formalize this observation.

First, can the low densities of *Eleusine coracana* be used to infer lower significance of this crop at this site? In other words is there any particular processing activity specific to *Eleusine coracana* that would explain these low densities and their inclusion into the archaeological record? *Eleusine coracana* can be regarded as a Type A or a Type B crop depending on whether it is irrigated or not. When it is well irrigated, the inflorescence heads are distinctly larger and can be selectively harvested as with Type A crops, however when the crop is not irrigated well the heads are smaller and the stalks are thinner and the most effective harvest is the Type II method. For the purpose of prehistoric interpretations, *Eleusine coracana* is regarded as a Type B crop. Unlike most Type B crops such as *Panicum miliare* and *Setaria italica*, *Eleusine coracana*'s caryopsis do not have binding glumes and husk. Instead the caryopsis are more like the naked seeds of *Pennisetum typhoides* and *Sorghum bicolor*. Therefore, the processing of *Eleusine coracana* does not require the roasting stage to help in dehusking. This stage is considered to be very important in the carbonization and inclusion of the seeds into the archaeological context.

The second clarification is related to the use of the plant crop. Would the use of the plant crop effect the recovery of its seeds in archaeological context? For example, if the plant was used only as green fodder (before the seeds were formed), then the importance of the plant to the community cannot be related to low frequencies of the seeds. In the case of the Babar Kot assemblage, the *Eleusine coracana* seeds do not represent a plant being used solely as fodder. If a plant was cultivated solely for animal fodder, then it is not allowed to mature and seed. Therefore, if *Eleusine coracana* was grown just for fodder, then it is unlikely that even 8 seeds would have been carbonized or recovered. Their very presence, albeit in significantly low densities, is an indication that their use was not limited and restricted to use as fodder. It is quite likely that the plant could have been used as fodder in addition to its use as human food. The low densities suggest lower usage as compared to the other millets recovered from the site. The presence of *Eleusine* sp. seeds can also be explained through two separate processes (if one assumes that the plant was not being used as fodder and food). The seeds of this plant could have been brought in as part of an assemblage of plants gathered for different purposes or were present in sod that was burnt as fuel (Pearsall 1988:100). I argue that even though the last two scenarios are plausible, it is more likely that the crop was being used as fodder and food. It is evident that the usage was limited (based on the significantly low density of seeds recovered at the site), and it appears to have been of secondary importance as a food plant.

Several contextual patterns observed in the Babar Kot archaeobotanical data need more detailed evaluation and specific application of ethnographic models, since the relevant sub-assemblages are very pertinent to the discernment of crop processing and crop usage at the site. These include the Occupation III seed pocket of *Setaria* sp. from the slope, the Occupation III trash pit on the slope, the Occupation III Harappan pot contents on the mound, an Occupation III ash refuse pit on the slope, two Occupation II pits on the mound, and an Occupation II clay feature on the mound.

Occupation III Seed Pocket, Slope

The composition of the seed pocket is most indicative of a lower crop processing stage. It is, however, important to note that the seed pocket contains two millet crop seeds (*Setaria italica* and *Panicum miliare*) and most likely includes a combination of crop processing activities performed on both the crops. The presence of *Setaria* rachii and spikelets with rachillas clearly indicates lower processing stages related to *Setaria italica* processing. The weed assemblage, however, shows a dominance of SFL weeds (66%) along with SFH weeds in strong frequencies (33%). This is indicative of a later series of the winnowing I process, a higher processing stage. If the weed assemblage is directly associated with the *Setaria italica* seeds, then there are two alternate scenarios for interpreting this crop sub-assemblage. The rachii fragments and spikelets with rachillas imply the presence of lower processing activities. The dominance of SFL weeds, however, suggests a later series of winnowing I. Therefore, the *Setaria italica* processing identified through the seed pocket composition includes lower and higher processing activities.

In contrast to *Setaria italica*, there are no *Panicum miliare* rachii or spikelets in the seed pocket. Negative evidence is used to argue for a different processing activity related to this crop, most likely a higher processing stage. It is, however, plausible that the some of the weeds (discussed for the *Setaria italica* case) are correlated with *Panicum miliare* processing. Therefore, one can argue for two scenarios in interpreting the *Panicum miliare* sub-assemblage of the seed pocket: either crop processing or food preparation. The assemblage may represent a very high processing stage such as winnowing II where only tail grains were being cleaned out and no weeds were present. Alternatively, the assemblage may represent a later series of winnowing I where there are no rachii, rachillas or spikelets, and only weeds. In either situation, the processing of *Panicum miliare* is more similar to a higher stage than a lower stage. Therefore, it is clear that the seed pocket reflects separate processing of the two crops.

Seed pocket seeds must have been burnt/charred through are intentional or accidental event, and several scenarios can be offered to explain this event. The first possibility is that the seeds were burnt during one of the crop processing stages, specifically the roasting stage which is done to aid in dehusking. Millets, which have tough glumes/husks, are very often roasted before the dehulling/dehusking process of pounding. The roasting aids in the efficient and successful removal of the hard clasping husk. It is possible that the *Setaria italica* seeds were accidentally spilled into the fire during a roasting process. Subsequent cleaning of the hearth and disposal into the domestic dumps resulted in their archaeological inclusion.

Another explanation is that the seed pocket contents are cleanings from the last few crop-processing stages (later series of winnowing I), which very typically occur in home bases and specifically in the food preparation areas. My ethnographic studies in India have shown that crop products during such a late stage of processing are very often cleaned of unwanted non-crop seeds and tail grains of the desired crop in the homes, often around hearths or in the yards. The cleanings are then most often swept into trash areas or if the cleaning was indoors then into the hearth. In the case of the Babar Kot seeds pocket, they could have been swept into a context where they were burnt, and then subsequently disposed into trash dump contexts.

A third possible explanation for the formation of this seed pocket is the inclusion in dung fuel as the result of being used as fodder. The millet grains could have been fed as fodder, and then subsequently the dung (with its undigested fodder seeds) was used as fuel in hearths for cooking. Later the cleanings from the hearths were dumped into the trash contexts. The last explanation is that they are the crop processing cleaning byproducts that were discarded into domestic areas and then were subsequently accidentally burnt, and entered into the dump context.

I argue that of these explanations the most probable is that the seed pocket represents byproduct cleanings from crop processing activities that were accidentally burnt either by being swept into the hearth or some other means. Subsequently they were deposited into trash contexts. For several reasons, the composition of the seed pocket is most indicative of such a process rather than one related to the roasting of *Setaria* sp. for dehusking. First, the seed pocket contains two crop seeds, those of *Setaria italica* and *Panicum miliare*. If only one crop plant were present, then it can be proposed that these seeds were being roasted for dehusking. However, since two crop seeds are present (albeit in different proportions), it is unlikely that they were both being roasted together for dehusking (i.e., mixed together). The clumped, aggregated and carbonized nature of the seeds negates the possibility that they were burnt separately and then subsequently deposited together. Therefore, the seeds of both crops had to have been carbonized together. The inclusion of two crop seeds into a context that was to be burnt later could have been two separate events, but they were burnt together. This therefore negates the possibility of the millets used as fodder and later the dung used as fuel, because if this scenario was correct then the millet seeds would not be clumped together.

The *Setaria italica* crop product cleaning was a deliberate and intensive selection against rachii, *Setaria* sp. spikelets with rachillas, and most probably also weeds. The byproduct cleanings were then swept into the burning context. But why were the *Setaria italica* crop seeds also being cleaned out? Most likely these seeds represent tail grains of *Setaria italica* that were selected out since their inclusion affects the quality of the final product and the food produced. The *Panicum miliare* seeds were either in that context (where they were disposed and got burnt with the *Setaria italica*) already through an earlier independent event or entered the context subsequently. Since no *Panicum miliare* spikelets with rachillas are present, it is argued that the same processing activity was not being performed on both the crops. It is likely that the *Panicum miliare* crop was being cleaned only for tail grains (i.e., the tail grains are often unsavory and could effect the quality of the ultimate food product).

The second reason for the seed pocket being more likely related to a lower crop processing stage (rather than a roasting accident), is the presence of rachii, *Setaria* sp. spikelets with rachillas, and weeds. These components would not be present at the stage of roasting for the purpose of dehusking. As demonstrated in chapter 4, by the roasting stage the only component left in the product is crop seeds. All other components have already been processed/ selected out. Therefore, one can strongly assert that the composition of the seed pocket is most indicative of a lower crop processing stage.

Occupation III Trash Pit, Slope

The trash pit shows series of distinct dumpings, with the top-most layers being more heterogeneous including seeds of *Setaria italica*, *Panicum miliare*, *Lens* sp., *Vicia* sp., and *Brassica* sp. The middle layers are characterized by *Setaria italica* seeds along with rachis fragments, *Linum* sp. seeds, and SFL weeds. The lower layers show a low density of carbonized seeds (0.2% of total assemblage of carbonized seeds from the feature) and comprise exclusively of *Setaria italica* seeds. The patterned variation in the dumping is indicative of several processes, the most important of which is that there is both variation and stability in plant discard (and therefore usage) at this locale of the site during the use-life of the trash pit. The patterning of plant seeds recovered, with legumes and *Panicum miliare* exclusively present in the topmost levels of the pit and *Linum* sp. seeds present only in the middle levels, is a strong indicator of the variation in plant discard and usage. In contrast, *Setaria italica* seeds are present through all the levels of the pit and therefore were being discarded (and used) throughout the duration of the trash pit usage. This trend is further evidence for varied importance of the different crops at the site, with *Setaria italica* being the key crop.

Another implication of the patterned variation of dumping into the trash pit is the presence of a distinctive assemblage only in the middle levels. This assemblage is most likely related to a cleaning byproduct of *Setaria italica*. *Setaria italica* rachis fragments are present only in the middle layers, and weeds are present in significantly high densities in these levels. The weed assemblage consists of SFL weeds (66%), SFH weeds (32%), SHH weeds (2%), and BFH weeds (0.3%). The dominance of the SFL weeds associated with the rachis fragments in this trash pit contexts demonstrates that this assemblage of *Setaria italica* seeds is representative of a byproduct related to a lower processing activity such as winnowing I or possibly sieving. These interpretations of the trash pit archaeobotanical assemblage in relationship to crop processing and food preparation are significant since they shed light on two important issues; the differential usage of plants at a specific locale on the site during one occupation time span, and the disposal of plant processing byproducts. Clearly, these plant byproducts were not fed as fodder to animals.

Two alternative interpretations of this assemblage as it relates to crop processing and/or food preparation are also possible: cleanings (or spillage) from crop processing stages or incorporation from fodder into the dung fuel. In both of these cases the seeds were carbonized before entering the pit context, most likely in a hearth (given the presence of burnt ceramics and ash), which was subsequently cleaned out and dumped into the trash pit. Of the two explanations the first is more plausible, mainly because of the high density of seeds recovered in the pit, which is unlike a dung fuel assemblage (Miller 1984). However, it is quite likely that crop-processing byproducts could have been fed as fodder and subsequently the dung from the animals (along with undigested seeds) was used as fuel. Therefore, all three explanations are feasible, that the assemblage is indicative of cleanings from crop processing or incorporation as fodder into dung that was later used as fuel. However, most importantly, the assemblage suggests the presence of lower processing stages on the site.

Occupation III Harappan Pot, Mound

Occupation III pot contents from within a room on the mound mainly included *Setaria italica* seeds of which a majority were broken/cracked. Two seeds of *Digitaria* sp., a SFH (Small Free Heavy) weed, were also present. It was established in the previous chapter that these seeds were indeed the original pot contents and not intrusives after the disposal of the pot. The contents are interpreted as being related to storage of *Setaria italica* after the dehusking/pounding stage. The dehusked state of the seeds is indicative that consumption was planned for not too distant future, and also that these particular seeds were not intended for trade or exchange. This reveals that a very high processing stage occurred at this particular locale, a habitation room at the site and unequivocally establishes that *Setaria italica* grains were used as human food. How the grain was consumed, cooked as rice or ground into flour, cannot be ascertained with the information available. Therefore, if one concedes that the seeds are truly the original pot contents, then the question of carbonization arises since in a majority of archaeological examples seeds stored in pots are not burned or carbonized and hence will not survive. Why were the seeds burnt/charred and then stored in the pot? *Setaria* sp. seeds have hard glumes and need to be dehusked before consumption. Often before dehusking, the crop seeds are roasted to loosen the husk and facilitate efficient dehusking. The storage of seeds is most effective (in terms of loss to pests and bacterial/fungal damage) before dehusking. One explanation for the carbonized seeds being stored in the pot is that the *Setaria italica* seeds were roasted for dehusking, and accidentally some were over charred. They were then dehusked through pounding activity and this is supported by the cracked nature of the seeds, and then subsequently stored in the pot for consumption. The storage of seeds after dehusking/pounding is a positive indicator that their consumption was planned for in the immediate future, and they were not being stored for the long term or for trade/exchange.

Occupation II Refuse Ash Pit, Slope Unit

The refuse ash pit is interpreted as being related to domestic cleanings from cooking fires. The archaeobotanical assemblage from this context is dominated by weed seeds (60%), with millets (35%)

and legumes (5%) occurring in significantly lower ratios. The weed assemblage consists of SFL and SFH weeds in varying proportions (41% and 59% respectively) and this is important for the interpretation of the pit's archaeobotanical assemblage as it relates to food preparation or crop processing activities. Since there are two millet crops and two legume crops in the pit, and there is no distinct pattern in the distribution of these various plants remains, it is not possible to correlate individual crops to specific crop processing activities or food preparation methods. Nonetheless, the weed categories are indicative of higher processing stages since the SFH weeds constitute the majority of the weeds (as discussed earlier, the SFH weeds are processed out later in the pathway since they are most similar morphologically to the crop seeds of *Setaria italica* and *Panicum miliare*). The relatively high frequencies of SFL weeds suggest a later series of winnowing I. Overall, the assemblage is indicative of later lower processing stages and higher processing activities. Considering the high percentage of the weeds, the assemblage is most probably the discard cleanings from a crop product through a form of winnowing, fine sieving, or hand sorting.

Occupation II Pits, Mound

The two feature pits on the mound in Occupation II are of importance because they are indicative of two different activities and related to two different crop plants, *Setaria italica* and *Panicum miliare*. One pit is dominated by *Panicum miliare* (75%) with SFH weeds in moderate ratios (25%). The other pit is dominated by *Setaria italica* (86%) and is more heterogeneous with SFH weeds (4%), SFL weeds (6%), *Lens* sp. (2%), and rachis fragments (2%). This patterned variation between the two pits is indicative of differential processing activities pertaining to the two crop plants. The first pit (dominated by *Panicum miliare*) is representative of the later processing series of winnowing I by shaking, since the SFH weeds with their morphology similar to crop grains are the last weeds to be cleaned out. The absence of rachis remains or spikelets with rachillas further supports this assessment. Ethnographic models indicate that often crop byproducts of the higher processing stages include crop seeds that are selected out because of their undesirable qualities (mainly size and fungal infection). Therefore, the contents of this pit are interpreted as byproducts from a relatively high crop processing stage of *Panicum miliare*.

The pit contents dominated by *Setaria italica* are interpreted as cleanings from a lower processing activity, such as the initial series of winnowing I. This assertion is primarily based on the presence of the rachis fragments, which are indicative of lower processing stages. The weed assemblage has a higher representation of the SFL weeds and the ethnographic models correlate assemblages dominated by SFL weeds to be most closely related to the later series of the lower processing stage of winnowing I. Thus, given these two lines of evidence, the pit contents were most likely related to winnowing I activities, and thus indicative of lower and a slight component of higher processing. In general, the contextual assemblages from the mound are indicative of higher processing activities, and to a lesser degree of lower processing activities.

Occupation II Clay Feature, Mound

The semi-spherical clay feature (which was interpreted as a hearth) has a heterogeneous assemblage with seven different components dominated by *Setaria italica* seeds (60%). The other components include a *Setaria* sp. rachis fragment, *Panicum miliare* seeds (5%), seed fragments of *Lens* sp. and *Lathyrus mungo* (.8%), and SFH weeds (2.4%). Unidentifiable seed fragments constitute an additional 30% of the assemblage. The presence of the four different crop plants implies that the assemblage is not formed by a single event but most likely is a composite of a series of events related to the individual crop plants. The usage of *Setaria italica* was clearly the most substantial but since there are several events represented, it is not possible to discern the activities related to processing of individual crops. However, given the very low ratios of weeds and rachii to the crop seeds it is clear that the crop products utilized at this locale were essentially cleaned products related to higher processing stages. Their deposition into this context could be a result of several processes including cleanings from later processing stages being swept into the hearth, their spillage during food preparation, and specifically in the case of *Setaria italica* and *Panicum miliare*, spillage during roasting for dehusking.

Archaeological patterns were interpreted using ethnographic models in the above discussion. Even though the archaeological patterns fit well to the expectations of the ethnographic models, it is important to address other factors that could produce such patterns as well. The main factor which could skew the interpretation of the Babar Kot archaeobotanical assemblage as being representative of lower and higher crop processing stages (and therefore indicative of cultivation and processing by the site inhabitants), is that the crops were being bought through trade and exchange as lower processing products or uncleaned higher processing products. These two situations are unlikely but still have to be examined and eliminated to strengthen the interpretations. If indeed the crops were being brought into the site as lower processing products, grown elsewhere, one would expect to find higher frequencies of big sized weeds, panicles, larger sized rachii, and rachillas. The Babar Kot archaeobotanical assemblage does not contain high densities of any of these components. Although it is possible that the site occupants could have brought in the crops as uncleaned higher processing stages, the probability is very low. Ethnographically there are only extremely rare situations when products from higher processing stages are unclean. Given the nature of Babar Kot, with its substantial architecture, sedentary nature and relatively high population, it is highly improbable that the occupants were obtaining crops through trade and exchange. Furthermore, the surrounding environs would have supported year round cultivation. Given this situation, it is argued that the Babar Kot occupants were undoubtedly cultivating the crops in the vicinity and this is reflected in the archaeobotanical assemblage of the site.

Economic Use of Plants: Food or Fodder

The Babar Kot assemblage can be interpreted as to being similar to an assemblage that is food-like, fodder-like or a combination of both. The botanical distinctions between food-like and fodder-like assemblages, as they relate to crop processing stages, have already been presented for Oriyo Timbo and the same distinctions apply for this discussion. It is necessary to comment that the fodder-like aspects of assemblages are the crop processing byproducts, but they need not necessarily be used as fodder. To ascertain whether these byproducts were used as animal fodder, the assemblages need to be from specific contexts (such as dung as fuel, dung plastering, etc.). In the absence of such contexts, then the fodder-like assemblages are considered crop byproducts that potentially could have been used as fodder.

Based on the patterns observed in the Babar Kot archaeobotanical assemblage, the entire assemblage is best correlated to being closely related to various crop processing stages' byproducts. Even though the assemblage is not food-like, there is definite evidence for use as human food based on the presence of crop byproducts from higher processing levels. Additionally, the presence of byproduct assemblages related to crop processing can be used to infer their use as fodder. However, additional archaeobotanical research at the site geared toward excavation and study of contexts where fodder-like assemblages would be likely recovered, such as hearths, floors and wall plasters, is required to strongly establish the presence of fodder-like assemblages. Equally important to this issue are other lines of evidence such as bone chemistry.

The millet assemblage patterning at the site shows a very distinct correlation to use as human food. Several contextual patterns have been interpreted as being related to lower processing and higher processing activities. These processing activities occur <u>only</u> if a crop is being used as human food. Any use of processed crop products as fodder is only secondary. In a majority of the contexts, the Babar Kot millet assemblage is indicative of the byproducts rather than products of crop processing, be it a lower processing stage or a higher processing stage. This negative mirroring of the processing is evidence for the use of millet crops as human food.

Model for Plant Usage and Subsistence at Babar Kot

The subsistence system at Babar Kot was geared toward cultivation of various crops, wild plant collection, and to an extent animal herding. The settlement is one of permanence with relatively substantial architecture and distinct domestic activity areas. Based on a variety of factors including the plants cultivated, the site appears to be occupied all year round, given the considerable size of the population, the subsistence system needed to accommodate long periods and relatively high demands. This was only possible through the practice of cultivation by the site occupants. It is very likely that wild plant gathering was also a common practice, and is indicated by the presence of *Ziziphus* sp. and other unidentified nut fragments.

The Babar Kot human subsistence system included the use of the millet grain, legumes, and oilseeds, supplemented by collected wild plants. It is important to reiterate the economic status of the different crops recovered from Babar Kot, especially the millets *Setaria italica*, *Panicum miliare,* and *Eleusine coracana*. Of the three millet crops recovered, *Setaria italica* was preferred most by the site occupants, followed by *Panicum miliare*. The millets as a group were the most important crops at Babar Kot and most probably were the basis of their subsistence system. They were most probably cultivated mainly in the summer monsoonal periods of the year. The two major millet crops (*Panicum miliare* and *Setaria italica*) recovered from Babar Kot appear to have been cultivated by the occupants in the vicinity of the site. Several non-crop plants in the assemblage (such as *Digitaria* sp., *Medicago* sp., *Phyllanthus* sp., *Trianthema* sp., and *Urochloa* sp.) are weeds that grow wild in millet fields and are good fodder for cattle.

Linum sp. was also a relatively important plant at the end of the Harappan occupation at Babar Kot (Occupation III). The oilseed was probably used for human food, and being suitable for cold and warm weather cultivation it added to the diverse nature of the farming system. Even though the leguminous crops such as *Lathyrus mungo*, *Lens* sp., and *Vicia* sp. were not very important as crops (as seen through their degree of recovery), their cultivation is significant since it further illustrates the diverse nature of the Babar Kot subsistence system. Two legumes, *Lathyrus mungo* and *Lens* sp., are winter cultivars and their presence is further indication that the Babar Kot occupants practiced year round agriculture.

Preliminary analysis of the faunal assemblage at Babar Kot reveals that animal meat (both domesticated and wild) consumption probably also contributed significantly to the diet of the inhabitants (Ryan 1991). Evidence of domesticated animals in the Harappan occupation of the site is strong and includes zebu cattle (*Bos indicus*), water buffalo (*Bubalus bubalis*), sheep (*Ovis orientalis*), goat (*Capra aegagrus*), pig (*Sus scrofa domesticus*), and dog (*Canis familiaris*) (Ryan 1991). A range of wild animal bones (probably hunted for food) was also recovered and include wild boar (*Sus scrofa cristatus*), nilgai (*Boselaphus tragocamelus*), chital or spotted deer (*Axis axis*), blackbuck (*Antilope cervicapra*), gazelle (*Gazella gazella*), and four-horned antelope (*Antilope cervicapra*) (Ryan 1991). Thus, an important component of the subsistence system was the use of wild resources (both plant and animal) by the occupants.

A model for plant usage at Babar Kot is presented in Figure 6-2. Elucidated in the model are the different processing stages and activities involved in the preparation of millet grain as human food: winnowing I, roasting, winnowing after roasting, pounding/dehusking and winnowing II. The disposal and filtering of the discard and selected seeds (both crop and non-crop) as relevant to this particular site are graphed to illustrate the complexity of the ultimate archaeological picture. Archaeological samples include discard as de facto, primary and secondary refuse in contexts of storage, roasting, cooking, and cleaning events. In contrast to the model of plant usage at Oriyo Timbo (Figure 6-1), the Babar Kot model has more activities discerned primarily because of the nature of the site and the character of the subsistence economy. As depicted in the Babar Kot plant usage model there are various plants brought into the settlement (crop plants, wild forage, collected wild plants for food, fuel and building material). However, the treatment of the crop plants is of primary concern, and therefore the model's emphasis is on these plants (Figure 6-2).

The model presents the interpretation that the inhabitants were bringing on site Type B millets (*Setaria italica* and *Panicum miliare*), legumes, and *Linum* oilseed as lower processing stage products (such as threshing product) and then completing the consecutive processing stages within the site. Each of the processing stages presented in the model was discerned and identified in different contexts at the site. Other processes, such as sweeping of discard and cleanings into hearths, trash, and inclusion into general sediments are inferred based on ethnographic research and the archaeological contextual patterns discerned at the site. The evidence for pounding/dehusking is strengthened by the recovery of grinding stones. These grinding stones were probably also used to process the grains into flour (after winnowing II process). The processing stages need not have necessarily been occurring in one locale, though there is distinct pattern of higher recovery of lower processing stage components from the slope area of the site. When a byproduct is swept into a hearth, and when it is discarded into a trash pit, is dependent on several factors; location of the processing activity, distance to the hearth, size of the hearth, time of last cleaning of the hearth, and amount of the byproduct to be disposed. Individual occupant variation is a very important factor in the relative role of each of the above factors. Similarly, when byproducts from the winnowing I series are given to animals as fodder, and when they are discarded, is also dependent on various factors. These include amount of byproduct, animal ownership, location

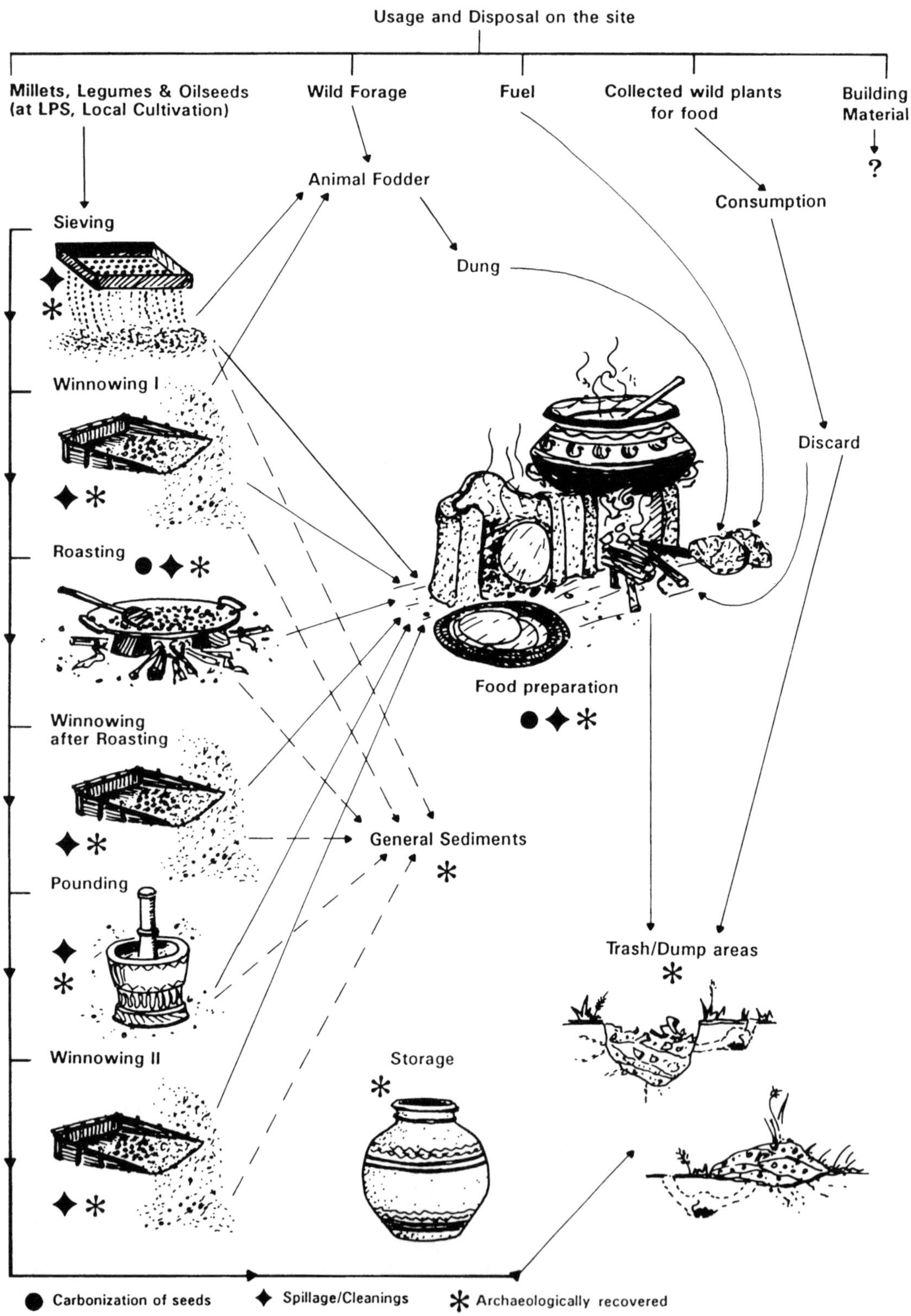

Fig. 6-2. Babar Kot plant usage model.

of processing activity, location/distance to animals, and lastly the subjective measure of necessity for fodder. There was a significant quantity of byproduct plant material generated through the processing of *Setaria italica* and *Panicum miliare*, and given the presence of domesticated animals at the site the byproducts were valuable animal feed, both as stored and immediate fodder. Millet byproducts are not an attractive fuel since they burn inefficiently.

The model presents the different plant utilization processes documented to have occurred at the site specifically for millet crops (Figure 6-2). However, it is important to note that all the processes must have been occurring each year, although how often a process occurs is uncertain. Nevertheless, the presence of these processes confirms the following observations: the status of millets as the primary food crops, oilseeds as secondary food crops, and the use of wild plants for fuel, food, forage and building materials. Thus, in contrast to the inhabitants of Oriyo Timbo, the Babar Kot inhabitants were self sufficient agriculturalists practicing year round farming of two (and perhaps three) main millets, oilseeds, and legumes, and also utilizing and exploiting the considerable wild resources available locally.

Summary of Archaeological Modeling

The ethnoarchaeological and archaeobotanical investigations at Oriyo Timbo and Babar Kot focused on addressing the economic role of millet cultivation for the occupants. The results reveal that the inhabitants of Oriyo Timbo were not cultivating millets at the site. Instead, they were bringing them in at the onset of occupation each season, or, were trading and exchanging them from neighboring agriculturalists. The character of and patterns in the archaeobotanical data indicate the use of millet grains primarily as human food. In contrast to this situation, the occupants at Babar Kot were involved in year-long local cultivation of crops, primarily millets, legumes, and oilseeds. The Babar Kot inhabitants were using millet grain as human food, and millets were the primary food crop for the settlement, and therefore an integral component of the economy. Oriyo Timbo archaeobotanical remains were mainly recovered from secondary refuse contexts such as general sediments and trash/dump pits, and to a lesser degree in primary refuse contexts such as hearths. The Babar Kot archaeobotanical remains were also recovered primarily from secondary refuse contexts (pits, trash dumps, and general sediments). However, de facto refuse in the form of pot contents and to a lesser degree primary refuse (hearth contents) were also recovered. Other differences between the two sites included the range of plant remains recovered from the three main cultural contexts. The Oriyo Timbo archaeobotanical assemblages from primary refuse contexts (hearths) show a significantly higher density of weed seeds, while Babar Kot primary refuse plant remains consist of higher densities of crop seeds with weed seeds accounting for a lower percentage. This significant distinction was used to argue for the varying use of the crop plants at the two sites.

Additionally, the recovery of secondary refuse plant remains from Babar Kot was higher than at Oriyo Timbo and this difference might be a reflection of length of settlement occupancy, occupancy numbers, and a higher number of processing activities at Babar Kot. The presence of de facto plant remains at Babar Kot is also important since it is indicative of storage. More extensive evidence of storage would indicate a form of long and short-term planning and also surplus production. However, these issues remain as speculation at this point. Thus, it is important to note the differences in refuse types, their compositions between the two sites, and their application for the respective subsistence economies.

The archaeobotanical and ethnoarchaeological investigations have successfully addressed the main issues of this study. The Oriyo Timbo subsistence economy was characterized by an absence of cultivation by the site occupants, while the Babar Kot subsistence system was defined through the identification of cultivation by the site occupants. Additionally, the use of millet crops as human food at both the sites was elucidated, as was the lack of millet use as fodder at Oriyo Timbo. The use of millets as fodder could not be definitively discerned at Babar Kot due to the complexity of the problem. It was, however, proposed that the likelihood of using millet byproducts as fodder by the occupants was very high given the amount of byproduct generated and the presence of domesticated animals. This hypothesis can be strengthened using other lines of evidence, such as bone chemistry and carbon isotope analysis of the domesticated animal remains from the site.

7. Modeling Animal Diet and Fodder Acquisition

This chapter explores new approaches for elucidating the role of fodder in domestic animal diet, specifically at the Harappan Phase sites of Babar Kot and Oriyo Timbo. Carbon isotope analysis is used to examine varied emphases in fodder, especially between wild forage and millets. Isotope analysis was done on wild forage plants, archaeological sediments, and animal bones. The procedure, problems and results of this analysis and subsequently, the ethnoarchaeological investigation of animal dung and its archaeological implications for fodder issues are discussed. In addition, ethnographic observations of forage acquisition and selection by pastoralists are presented. The chapter concludes by presenting an animal feeding model for the Harappan Phase in Gujarat.

Carbon Isotope Analysis and Results

The relative importance of millets (byproducts or green crops) as animal fodder can be examined through carbon isotope analysis of animal bones. Ratios of the two stable isotopes of carbon, Carbon 13 (^{13}C) and Carbon 12 (^{12}C) in animal bone collagen have been used to determine the relative contribution of C4 plants (such as maize and millets) versus C3 plants (such as wheat and barley) to the diet of the consumer (DeNiro and Epstein 1978; van der Merwe 1982). Millets (such as *Eleucine coracana*, *Pennisetum typhoides*, *Panicum miliare*, *Sorghum bicolor*, and *Setaria italica*) are C4 plants, and if they were the major source of animal fodder, then animals with a diet consisting of a high proportion of plants with C4 photosynthetic pathways should be reflected in a relatively heavy carbon isotope signature of the animal bone collagen (DeNiro and Epstein 1978). The use of 'heavier' to indicate a less negative $\delta^{13}C$ value (richer in $\delta^{13}C$) and 'lighter' to indicate a more negative $\delta^{13}C$ value (less rich in $\delta^{13}C$) follows the practice of O'Leary (1981:553). Generally accepted $\delta^{13}C$ ranges for C3 plants are -20% to -35%, while the $\delta^{13}C$ ranges for C4 plants are -9% to -16% (van der Merwe 1982). Provided that well-preserved bone is available, carbon isotope analysis is ideally suited to determine the extent of millet use during the Harappan Phase, because millets were the primary C4 plants of economic importance in Gujarat.

In general, with respect to the study of Harappan subsistence economy in Gujarat, the exact proportion of millets in animal diets, particularly cattle, is not essential. Instead, the relative long-term changes in the dependence on the different fodders, particularly millets versus wild forage, are of key importance. An increased emphasis on millets as fodder during the Late Harappan of Gujarat may indicate the emergence of a specialized subsistence system, with animal husbandry as an important and critical variable. This has parallels to present-day Gujarat (Reddy 1991a). Such a change in subsistence economics, entailing specialized food production and intensification, could have far reaching implications for the Late Harappan period of Gujarat. Increased emphasis on fodder cultivation implies more specialized subsistence procurement and a shift from a previous non-specialized procurement system. This research proposed to investigate this inference through the application of carbon isotope analysis of animal bones (specifically cattle) from Babar Kot and Oriyo Timbo, in order to complement the insights gained from the ethnographic crop processing studies and archaeobotanical analysis of plant remains from two Harappan sites.

In order to discern the relative importance of millets as fodder versus other wild plants, it was necessary to ascertain two variables: the nature of the environment, and whether other C4 plants were present which could skew the millet signatures. Soil samples from the different occupation strata at Oriyo Timbo and Babar Kot were chemically analyzed for soil organic carbon isotope signatures to ascertain the nature of the environment at the time of occupation. It is important to determine whether a C3 or a C4 environment existed locally because this would help elucidate whether C4 plants were dominant. This observation is crucial in establishing millets as primary plants of economic importance. In other words, for this technique to be useful, the local environment should have a different carbon isotope signature than the crops under investigation. Typically, C4 environments are indicative of lower rainfall and arid conditions, while a C3 environment is more indicative of temperate climates. All of the laboratory analyses of soils, plants, and bones related to carbon isotopes were conducted by the Paleodiet Laboratory, University of Wisconsin, Madison, under the supervision of Dr. Margaret Schoeninger.

A total of nine sediment samples were submitted for analysis, four from Babar Kot and five from Oriyo Timbo (Table 7-1). The $\delta^{13}C$ values of the soil organics in the Oriyo Timbo Harappan levels ranged -18.2% to -22.6%, with an average of -20.4%. This lighter $\delta^{13}C$ value is indicative of a strong C3 signal, thus implying a higher ratio of C3 plants as compared to C4 plants in the environment during the site occupation. The Babar Kot soil organics show similar lighter signatures with the $\delta^{13}C$ values ranging from -17.5% to -21.2% in the different strata, with an average of -19.4%. This again indicates a higher percentage of C3 plants in the environment at the time of occupation. Therefore, the soils from the occupation levels of both sites sampled in this project show a strong C3 environmental signal, and thus any C4 signature derived from animal bones cannot be attributed to the local environment but rather to specialized feeding.

Along with soil samples, ethnographic samples of wild fodders were collected and analyzed for their carbon isotope ratios. These fodders were specifically analyzed because they are currently selectively chosen by pastoralists and farmers as cattle fodder. A total of 19 different wild fodders were analyzed for their $\delta^{13}C$ values. Approximately 70% of them were C3 plants, and 30% were C4 plants. The $\delta^{13}C$ values for the C3 plants ranged from -23.9% to -30.5%, and the $\delta^{13}C$ values for the C4 plants ranged from -10.1% to -12.9%. Some of the C4 plants such as *Trianthema* sp., *Amaranthus* sp., and *Setaria tomentosa* are also prominent field crop weeds. This predominance of C3 plants ethnographically is also corroborating evidence for a local C3 environment in the past since the environment is believed to have been similar to the present specifically during the Harappan Phase.

The $\delta^{13}C$ values for soils and wild fodders make the interpretation of the animal bone collagen isotopic ratios relatively straightforward, since there would be limited confusion from other sources of C4 fodders apart from millets. However, if cattle ate considerably more C3 plants than the C4 millets, then the millet signature would be obscured. Thus, one needs to consider the seasonality of fodder availability and the mixed feeding on C3 and C4 plants in such an investigation. In addition to bone preservation in the local sediments and post depositional changes, the chemistry of the animal bone collagen is ultimately dependent on a number of variables. These include the nature of local sediments and vegetation, herding characteristics, nutritive value of forage, forage availability, selection, and potential of fodder intake. For archaeological application, only some of these variables can be observed and quantified through ethnographic studies. These include seasonality of fodder, forage nutritive quality, and selection by animal and herder.

My ethnographic studies show distinct seasonality in the availability and consumption of fodder by cattle and goats. As just discussed, such selected fodder (for their nutritive quality) was analyzed for their $\delta^{13}C$ values and a significant majority are C3 plants. These plants are, however, seasonally available, mostly in the winter months. For example, as presented in the earlier chapter discussion of the Oriyo Timbo subsistence system, often wild fodders such as *Cressa cretica* (a C3 plant) are preferred over millets and other crops or their byproducts. *Cressa cretica* is readily available in the winter and enhances milk production in cattle and goats because its high protein content induces lactation. It is used fresh, and when available it is given primary preference, since it cannot be stored as dry fodder.

Carbon isotope analysis cannot recognize seasonality in fodder consumption, because of the fact that bone collagen has a slow turnover rate. But what is of primary concern for questions related to economic importance of millets to the occupants of Oriyo Timbo and Babar Kot is the relative propor-

Table 7-1. Soil Organics of Oriyo Timbo and Babar Kot Soil Samples.

SITE	OCCUPATION	$\delta^{13}C$ (°°/°)
Babar Kot	III	-21.2
Babar Kot	III	-20.2
Babar Kot	II	-19.6
Babar Kot	II	-17.5
Oriyo Timbo	II	-18.2
Oriyo Timbo	II	-21.4
Oriyo Timbo	II	-22.6
Oriyo Timbo	I	-18.4
Oriyo Timbo	I	-18.4

tion of millets versus the wild C3 plants in the animal diet rather than the exact amount. Using the data from the wild fodders, the $\delta^{13}C$ values of archaeological animal bones from the two sites can be interpreted as being indicative of higher proportions of a C3 versus C4 diet, or a mix of the two. Although carbon isotope analysis has shortcomings regarding the detection of seasonality and other related processes, the signatures will provide insights into the long term proportional feeding of C3 and C4 plants.

A total of 22 cattle bones were submitted for carbon isotope analysis, 11 from Babar Kot and 12 from Oriyo Timbo. Unfortunately, except for two samples, the carbon isotope analysis of faunal remains from both the sites was unsuccessful because of the poor condition and preservation of collagen in the faunal bones. Two different methods of collagen retrieval were used. The first was the HCL method of extraction using two different techniques; first as done by DeNiro and Epstein (1978, 1981) and later improved by Chisholm et al. (1983) and Schoeninger and DeNiro (1984). The other HCL method used was developed by Sealy and van der Merwe (1986). When neither method was successful, extraction through the EDTA demineralization procedure (Tuross et al. 1988) was employed because they have shown that the yield of collagen obtained through EDTA demineralization was consistently higher than extraction procedures that used HCL. The EDTA extraction method is particularly appropriate for fossil samples with small amounts of organic material (Schwarcz and Schoeninger 1991). Despite these intensive retrieval procedures, collagen was extracted only from two cattle bones, both from Oriyo Timbo strata 6b of occupation I. The $\delta^{13}C$ values of the collagen from these bones were -30.063% and -27.790%, both strong C3 signals. The soil $\delta^{13}C$ signature for occupation I is -18.4% also a C3 signal. Of concern are the high C:N ratios of 10 and 8.2 obtained for the two bone samples. The accuracy of the $\delta^{13}C$ signatures of the bones depends on whether the $\delta^{13}C$ values measured intact collagen protein or proteins with amino acids compositions similar to the intact proteins (Schwarcz and Schoeninger 1991). The use of C:N ratios in the past as a screening procedure (DeNiro 1985; DeNiro and Epstein 1981; Schoeninger and DeNiro 1984) has been proven ineffective. According to Schwarcz and Schoeninger (1991:294), the C:N ratios outside the acceptable range of 2.7 - 3.6 is indicative of noncollagenous material, however a C:N ratio within the range does not suggest acceptable collagen either. Instead they argue that organic residues should weigh at least 5% of the original bone weight to use the C:N ratios as indicative of non-collagenous material. In the case of the two Oriyo Timbo bones with the ^{13}C values, the organic residues weights were less than 5% of the original bone weight. Therefore, the C:N ratios cannot be indicative of collagenous material and the $\delta^{13}C$ values determined for the two Oriyo Timbo bones must be considered inconclusive.

Nonetheless, even though specific isotopic data are lacking, it is useful to discuss the theoretical implications of various isotopic signatures for understanding the use of different plants as fodder and to reconstruct animal diet during the Harappan Phase in Gujarat. This model can then be applied at sites where collagen preservation is better.

Predictive Modeling of Alternate Fodder Emphases

It is possible to predict the subsistence systems in play if indeed there was a C3 or a C4 signature determined for the Oriyo Timbo and Babar Kot domesticated animal bones in spite of the failure of collagen extraction and the lack of carbon isotopic data. The two main situations would be either the elucidation of a relatively heavy $\delta^{13}C$ signature (indicating a C44 plant diet) or a lighter $\delta^{13}C$ signature (indicating a C3 diet),

A relatively heavy $\delta^{13}C$ signature from the cattle bones would imply that C4 plants were consumed in greater proportions. Millets were probably a major contribution to the animal's diet since the majority of wild fodders are C3 plants and their inclusion into the animal's diet would not confuse the issue. If indeed a heavy $\delta^{13}C$ signature has been obtained, this being indicative of higher consumption of millets, a critical issue would be whether millet grains or millet byproducts were being used as animal fodder.

This distinction is significant since the use of millet grain for animal fodder would indicate that the crops were primarily grown for animals, whereas using millet residues for fodder implies that the crops were grown for both human and animal consumption. The difference has a significant effect on the interpretation of the subsistence economy. Although it may seem counterproductive to cultivate millets only for animal fodder, ethnographic data from traditional, non-mechanized farmers and pastoralists show that, very often, quick maturing millets are cultivated solely for animals as insurance against lean periods (Purseglove 1979). However, it is important to note that when millets are cultivated solely for animal fodder, they are not allowed to seed, since fresh green fodder is more effective in terms of its nutritive quality, palatability, and the digestion and utilization of nutrients. Therefore, when millets are cultivated for animal fodder, the crop is harvested green and fed directly to the animals. In the rare cases when the crop is allowed to seed, the crop is still cut and fed without any processing. In areas of extreme seasonality of fodder availability, the harvested millets crop plants are stored for usage during the lean months. In either case the millets cultivated solely for animal fodder are never processed through the various stages described in chapter 4. This is a very important distinction, and can be used in archaeological contexts to determine whether a crop is cultivated for human food or animal fodder.

Oriyo Timbo and Babar Kot both show a high recovery rate of millet crop seeds from archaeologically distinct contexts related to domestic activities. The presence of seeds in such contexts is therefore indicative of usage primarily as human food since the cleaning of crop products through an intensive processing pathway would not take place if the only intention was use as animal fodder. Additionally, since Oriyo Timbo inhabitants do not appear to have been cultivating millets at the site as determined through the ethnoarchaeological modeling, their access to millets as animal fodder would have been minimal. Therefore, a $\delta^{13}C$ value with a C4 signature would indicate predominant selective feeding of C4 wild fodders. Since the soils indicate a C3 environment, such a result implies that the Oriyo Timbo inhabitants were controlling and purposely selecting the diet of their animals.

A $\delta^{13}C$ value for the animal bones indicative of a C4 diet would be easier to explain for Babar Kot than Oriyo Timbo. Crop byproducts at the various processing stages are viable animal fodder, and in the case of Babar Kot where millets were being cultivated by the site occupants (as opposed to Oriyo Timbo inhabitants who appear to have been obtaining the millets through trade and exchange) it is very likely that the crop byproducts from the different processing stages were used as fodder (both immediate and stored) for the domestic animals such as cattle. Several contexts were identified at Babar Kot (and discussed in chapter 5) that are indicative of such processing byproducts. However, one might ask if these assemblages are indicative of cleanings from these processing stages, or if they are indicative of use of the byproducts from these processing stages as fodder, and subsequently included in animal dung used as fuel? The distinction might seem significant, but, for the questions posed in this book, the distinction is not as critical for two reasons. First, to identify cultivation by the inhabitants, the presence of assemblages indicative of lower processing stages (and therefore local cultivation) is in itself sufficient regardless of their context (in dung or as distinct cleanings). Secondly, for the question of use of millets (grain or byproducts) as animal fodder, the presence of the crop processing byproducts in a dung-related context strongly supports the hypothesis of fodder use. However, even if dung contexts are absent, the given presence of the byproducts increases the likelihood for their having been used as fodder.

The implications of a lighter $\delta^{13}C$ (C3) signature are equally important and have far reaching inferences, particularly with respect to the 'pastoral' character of Oriyo Timbo and the importance of animal husbandry to the Babar Kot occupants. Since the wild fodders in the area appear to be mostly C3 plants, one can infer that such a pattern is the result of the animal diet being primarily based on wild forage. In the case of Oriyo Timbo, such an emphasis would fit well into their subsistence economy, which was based on acquisition of human food grains (such as millets) through trade and exchange. In this situation the Oriyo Timbo inhabitants obtained the primary food grains from outside, and raised their animals on wild forage (through intentional collection and/or free range grazing). In case of Babar Kot, a $\delta^{13}C$ signature indicating that animal diet was primarily based on C3 plants would be harder to explain. Cultivation of millet crops by the inhabitants would have no doubt generated a considerable quantity of byproduct (from crop processing), and their lack of use as fodder would need to be explained. The deliberate selection against the millet byproducts is unlikely. However, factors that could have been important are seasonal availability of millet byproducts and the quantities available. In such a case, the reliable ratio of C3: C4 plants in the animal diet would need to be examined further.

Nevertheless, if the animals were feeding primarily on wild fodder, then the question arises whether a sizable herd really could be maintained in the environs around Oriyo Timbo and Babar Kot. This would be particularly problematic during the lean dry periods of the year. Therefore, in the case of Babar Kot, herds may have been taken to other locales during the dry season. This solution is not applicable for the Oriyo Timbo inhabitants since they probably occupied the site during the dry season. Therefore, they must have survived on local fodder acquisition. These issues can be examined only when further conclusive carbon isotope data on animal bones are available.

Animal Dung as Fuel in Domestic Hearths

In a seminal study of dung fueled fire contexts in modern Iran, Miller (1984) stressed that dung used as fuel can be a major source of seeds in fireplaces, and suggested that the carbonized seeds recovered from such contexts are probably indicative of animal fodder rather than human food remains. Bottema (1984), Miller (1984) and Miller and Smart (1984) have found carbonized and uncarbonized seeds in modern dung fueled fire contexts. Miller (1984) determined that seeds found in hearth samples could have come from several sources including food residues spat into fires, intentionally discarded debris from the cleaning of grains and crops, food processing near fires, cooking accidents, ambient weed seeds blown or dropped into a fire, and dung burned as fuel. She found that of all these sources, dung was the major source of seeds in the hearth contexts. The studies cited by Miller and Smart (1984) on the effect of animal digestion indicate that seeds were found to be present in cattle dung for at least 10 days after they were ingested. Various animals were tested, including cattle, sheep, chickens, hogs, and horses. Miller and Smart (1984) concluded that seeds were present in nearly all the dung samples but the numbers of seeds recovered varied in terms of species and often the time lapse since ingestion affected the number of seeds recovered. Bottema's (1984) study of modern fireplaces in Syria revealed that a majority of seeds recovered from these contexts are a result of the use of dung as fuel. His research also involved the study of dung cakes before use as fuel, and comparison between the contents of dung fueled hearths and the dung cakes. Bottema (1984) determined that although seeds could make their way into hearths from extant vegetation under the hearth-fire, ambient seed rain, grains spilled through cooking, and vegetable fuels, the majority of the seeds were from burned sheep dung.

Although these results have profound implications, concerns have been raised as to whether seeds in dung would survive high temperatures in hearths (Hillman 1984; Reddy 1991a, 1994). Certainly, Bottema (1984) and Miller (1984) present viable models for archaeological contexts, given the recovery of dung fuel remnants and high concentrations of charcoal (as argued by Miller[1984] for dung fuel contexts). Bottema (1984) also acknowledged that several grass seeds preserved in unburned dung disappeared through charring in the hearths.

Very importantly, these previous studies in the Near East do not specifically address the use of crop processing byproducts as fodder, nor the issue of distinguishing the ingested seeds from the inclusion of byproducts into the dung as temper, or their inclusion into the hearths by disposal during processing and cleaning. The latter two situations have recently been observed in ethnographic contexts in India during this study. Miller and Smart (1984) and Miller (1984) discuss and acknowledge potential multiple sources of seeds in hearths, and do not suggest that there is a single source for all seeds. Miller (1984) recognizes that contexts where crop processing residues are tossed into dung fueled fires as trash result in assemblages that are impossible to distinguish from assemblages created by inclusion of seeds into dung as fodder and subsequent use of the dung as fuel. In addition, Charles (1990) presents ethnographic observations from southern Iraq on the role of animal diet and addition of crop processing residues in dung packing. However, none of these studies define nor discern the range of crop processing activities that occur within the vicinity of the hearth, such as crop product cleaning, which could directly affect the composition of hearth contents.

Since the studies of Bottema (1984) and Miller (1984) were in the Near East, an ethnographic pilot study of dung samples and hearth samples was conducted in the modern state of Gujarat, northwest India to examine regional patterns and address specific questions related to hearth contents and food processing activities carried out adjacent to hearths. The objective of the study was to gain insights into the use of traditional fuels, particularly dung, in domestic hearths and its implications for archaeological interpretations of carbonized plant remains from analogous contexts, and the formation of hearth contents. The study included a combination of interviewing, mapping of hearths, and collection and analysis of hearth and dung contents.

This study, built on previous work by Miller (1984), argues that a number of other factors can also contribute to the presence of seed in dung fueled hearths, by examining the composition of dung cakes prior to use as fuel, and ash samples from dung fueled hearths. The study included a combination of interviewing, mapping of hearths, and collection and analysis of hearth contents. A range of hearths was studied for insight into their construction, including make-shift hearths in the field, make-shift domestic hearths, stationary domestic dung plastered hearths, and mobile dung plastered hearths. The composition analysis of dung and dung-fueled hearths from India clearly indicates that it is not always possible to conclusively distinguish the source of carbonized seeds from hearth contexts (see Reddy 1999 for further details). They may originate from the use of dung as fuel, wherein specific plant seeds get included in the dung through animal consumption; they may result from various plant materials which are purposely or accidentally included in the dung as temper; or they may come from the disposal of crop byproduct cleanings into the hearths (note that dung temper refers to the vegetal materials used to temper dung, and not dung functioning as a tempering material).

Ethnographic cattle dung cake samples were analyzed for their composition. The dung cakes were collected from the yard of a traditional farming group within a settled community. The samples were comprised of cattle dung along with millet crop straw used as temper, some non-crop vegetative matter, seeds, and over two dozen goat dung pellets also used as temper. A total of four dung cake samples were analyzed (Table 7-2). The seeds included millets, legumes, and a range of weed seeds. The cultigens accounted for 2.6%, while weed seeds comprised 97.4% of the plant seeds recovered from the dung cakes. The dung cakes analyzed in this study are similar to Miller's (1984) dung samples in terms of the relative percentage of different plant categories. Three ash samples from hearths where dung was used as fuel were also analyzed. The seeds recovered from these contexts included millets, legumes, and weeds, with cultigens accounting for 64% and weeds accounting for 36%. These frequencies differ substantially from Miller's (1984) ethnographic assemblages from dung-fueled contexts where a significantly higher percentage of weed seeds was recovered (87 gm of sample yielded 81 wild and weedy seeds and 1 cultigen seed; see Miller [1984]).

Thus, in contrast to Miller's (1984) findings, there is a significant distinction in the composition of the dung cakes and dung fueled hearth samples in this study. The two major differences between the dung cakes and ash samples are botanical composition and density of seeds recovered per liter. With respect to the botanical composition of the two assemblages, cultigens comprise the majority of the ash assemblage from the hearth, while they occur in very low densities in the dung assemblage. Weed seeds, however, occur in higher densities in the dung samples but in significantly lower densities in the ashy hearth sample (see Table 7-2).

The reversed ratios of botanical composition are significant for archaeobotanical interpretations regarding the origin of seeds (either from dung or from crop processing byproduct disposal). It is possible that the overall lower frequencies in the hearth samples are the result of the destruction of seeds (both cultigens and weeds) from the dung cakes during the fueling process. The high frequency of cultigens in the hearth samples is most probably a result of their discard into the hearth during the adjacent final cleaning of the crop product. Ethnographic research in India has revealed that there is a high incidence of crop processing disposal into the hearths during the final cleaning of crop products for food preparation and immediately before food preparation. Crop products in the last stages of processing very often are cleaned of unwanted tail grains of the desired crop in the homes, often around hearths, just prior to food preparation, or in the yards. Tail grains of millets are typically undesirable due to their small size and poor, often bitter, taste. The cleanings, if outside, are then most often swept into trash areas; if the cleaning takes place indoors, the byproduct is disposed into the hearth. Thus, in these samples the higher frequency of cultigens as compared to weeds in the hearth samples is possibly due to nearby domestic activities and not due to dung fueling of the hearth.

Table 7-2. Compositions of Ethnographic Dung and Ash Samples.

Sample	**Sample Volume (liters)**	**Total Seeds (N)**	**Seeds/liter (N)**	**Cultigens/liter (N)**	**Cultigens/liter %**	**Weeds/liter (N)**	**Weeds/liter %**
Dung Cake	10	1161	116	3	2.6%	113	97.4 %
Hearth Ash Sample	10	253	25	16	64%	9	36.0 %

The assemblages from the various contexts at Babar Kot are indicative of both the situations (i.e., some contexts have high percentage of weeds while others have high percentage of millets), and I argue that it is not possible to conclusively distinguish the use of byproducts as fodder (and subsequent use of dung as fuel) and the disposal of byproduct cleaning into various contexts. As stressed earlier, the lack of resolution is not critical, since the presence of the byproducts demonstrates the presence of certain processing stages and through ethnographic observation it is proposed that there is a high probability these byproducts were used as fodder. Additionally, the status of *Eleucine coracana* at Babar Kot as a primary human crop is questioned, due to its occurrence in very low frequencies. It is hypothesized in this study that the plant crop might have been cultivated solely as fodder and this may account for the lack of evidence for crop processing stages and the low seed frequencies. Thus, the Babar Kot inhabitants appear to have cultivated *Setaria italica* and *Panicum miliare* as human food, while *Eleucine coracana* was possibly only a fodder crop. The choice of specific crops for different purposes would have been dependent on several factors, including necessity and the success of cultivation.

Animal Feeding Model for Gujarat Harappan

Based on the available information (ethnographic, archaeological, and theoretical isotopic signatures) a predictive animal feeding model for the Harappan Phase in Gujarat is proposed (Figure 7-1). The model is specifically intended for domesticated herbivores, such as cattle, pigs, goats and sheep, all of which are present at the sites of Oriyo Timbo and Babar Kot.

The animal feeding predictive model can be employed when isotopic data from animal bones are available, and at that time the various implications of the isotopic data can be interpreted. For example, if a C3 signature is obtained for the animal bones, the model can be used to argue for specific selective and natural processes affecting the diet, which ultimately produced the specific signature for the animal. Since it was intended for this purpose, the dietary effects on animal collagen are the major focus of the model. Even though the animals considered in the model are domesticated, their feeding behavior is not random, instead it is a highly selective and distinctive activity.

The model presents six different factors that affect fodder choice: whether the animal is going to feed primarily on millets (green crop or byproducts) or wild forage (collected or free range) (Figure 7-1). The factors are level of intake, herd effects, potential intake, nutritive value, selection, and economy. The relative importance of each factor is determined by several conditions. The level of intake is primarily affected by the quality (and therefore the palatability) of the available fodder (fresh or dried), and also the seasonality in the availability of different fodders. Potential intake is an animal dependent variable, indicating that the physical condition of the animal affects the choice of fodder. Nutritive value of the fodder can be seen as a determining factor in the choice particularly for the wild forages. Caswell et al. (1973) conducted a study of herbivore diet selectivity and photosynthetic pathways, and proposed that the C4 wild plants in general have a lower digestibility, palatability, and nutrient value due to inherent photosynthetic processing locations. This implies that digestibility and effective utilization of nutrients in herbivores are higher for C3 plants than C4 plants (Caswell et al. 1973; Wilson and Ford 1971). Their study was conducted on grasshoppers, and it is not certain whether the results are applicable to domesticated animals, specifically ruminants like cattle. They also proposed that herbivores avoid C4 plants rather than prefer C3 plants. If this assertion is correct, then the nutritive value choice of wild forage would be selectively biased toward C3 plants.

Herd effects also determine the mix or dominance of millets versus wild forage in the animal diet. The main herd effects of concern are grazing and browsing. Browsing, which entails feeding on scanty vegetation, is very common in Gujarat. However, browse vegetation is usually fibrous and contains tannins (essentially oils and other aromatic compounds) that decrease palatability. In addition, browse plant species are of low palatability and eaten primarily in the dry season when herbaceous forage becomes either sparse or unpalatable. Grazing entails eating grasses and woody materials. Animal species differ in their preference for browsing and grazing, with sheep and cattle being primarily grazers, and eating little of the browse type plants. It is quite likely, however, that the zebu cattle graze to a lesser extent. Goats are often considered intermediate feeders, with browsing being the most common feeding method combined with occasional grazing. There is seasonality in the digestibility of herbaceous grazing forages but not in browsing, which is an important distinction to consider in the model.

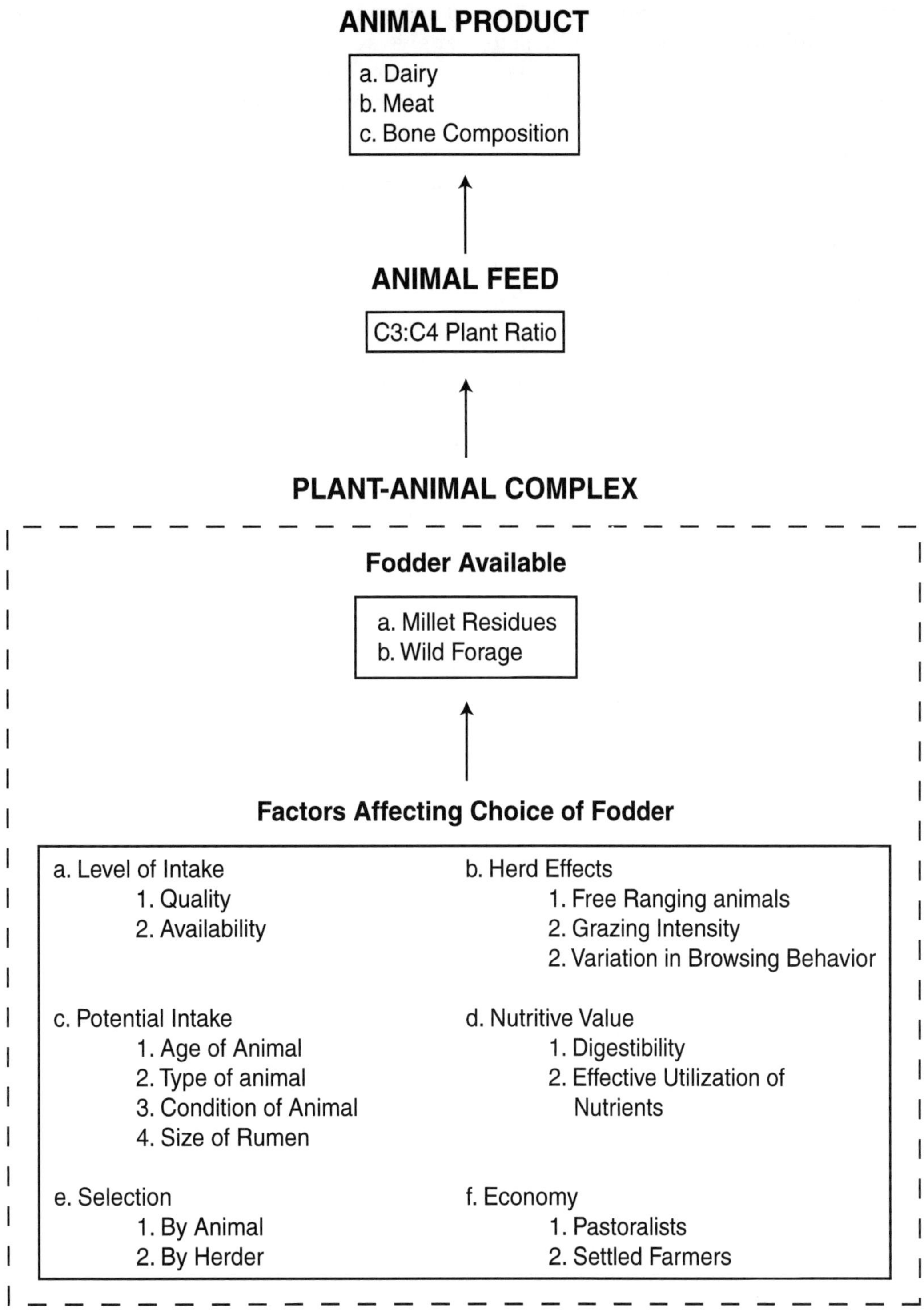

Fig. 7-1. Animal feeding model.

Conscious fodder selection by the animals and the herders is another important factor directly affecting whether or not millets or wild forage are dominant in the animal diet. For example, there is selective fodder acquisition by pastoralists, dependent on the effect the fodder has on milk production. The ethnographic and isotopic studies have determined that the wild forage collected by pastoralists shows a clear dominance of C3 plants. Diet selection by the animal is fundamentally a behavioral process that constitutes the interface between the free-ranging animal and the plant community upon which it feeds (Heath et. al. 1985). While the existence of linkages between diet selection and nutritional status of the animal are obvious, the components and dynamics of these linkages are exceedingly complex and inadequately understood. In spite of the research efforts there is little known about the predictability of nutritive composition and diet selection in a given set of circumstances (Health et al. 1985). Nevertheless, diet selection (on the part of the animal and the herder) is a recognized behavior that needs to be considered when determining which plant groups (millets or wild plants) dominate the animal diet.

The last factor that influences the choice of millets or wild plants in the animal diet is the subsistence economy of the social group whose animals are being considered. For a community involved in farming and cultivation of millets on a regular basis (such as at Babar Kot), millet byproducts as fodder are readily available. It is highly probable that animals are frequently fed on the residues, particularly in months when wild forage is scarce or free range grazing is not viable. Additionally, some millets might be selectively grown only as green fodder. In such a context, determining the proportion of millets and their residues in the animal diet is of critical importance. This can be answered only through carbon isotope analysis of the animal bones. In contrast, a seasonally mobile nomadic pastoralist community (such as at Oriyo Timbo) focused on animal husbandry and involved in the procurement of millets from outside primarily for human consumption is unlikely to have millet residues available on a regular basis to feed the animals as fodder. Whether or not millets (green fodder or residues) are obtained for fodder from non-local sources is dependent on several factors such as the season of occupation (since millets are summer cultivars) and the availability of wild forage. Again, the critical question regarding what proportion of the animal's diet are millets (green fodder or residues) can only be determined through carbon isotope analysis.

The various components of the animal-plant complex in the model (Figure 7-1) directly affect the C3:C4 animal feed ratio. The plant ratio is likely to change seasonally, and over the life-span of the animal, and since collagen has a slow turnover rate what is discernible in the bone collagen are the diet ratios in the last years of the animals diet. When carbon isotope data are available for the Gujarat Harappan sites, the model can be put to effective use to interpret the significance of the signatures. In summary, it is proposed that the diets of the Oriyo Timbo domesticated animals were primarily comprised of wild forage, while the Babar Kot domesticated animals were fed millet byproducts when available (November onwards), and wild forage supplemented their diet during other seasons.

8. Conclusion: Discerning Palates, Plant Usage and Subsistence

Initially, I posed several questions regarding the economic role of crops and the importance of summer crops during the Harappan Phase in Gujarat, India. To address these issues a new methodology was developed tailored to identify crop cultivation in archaeological contexts as a distinct activity and also to elucidate varied roles of millet cultivation in different types of sites including sedentary settlements and ephemeral camps. In particular, since artifactual ties between Harappan pastoral and sedentary settlements in Gujarat are absent (or yet to be identified), it was suggested that economic indicators, such as presence or absence of cultivation, provide the best line of evidence to understand the relative degree of integration between pastoralism and agriculture. If a tightly integrated system was in play a broad range of economic choices was possible. Settlements might be occupied by pastoral nomads or agriculturalists, or by populations practicing subsistence strategies that fall between these two extremes (such as semi-nomadic or semi-sedentary pastoralists). Variants of these systems, such as cultivators and pastoral nomads complementing each other or cultivators also being pastoralists are important possible adaptations to consider in the elucidation of subsistence and settlement systems. The key issue would be to understand how and to what degree particular cultivated products might fit into pastoral adaptations (Khazanov 1984).

Prior to this study, cultivation of the millet crops as a distinct economic activity had not been clearly demonstrated at any Harappan site. The identification of actual cultivation is important in order to exclude the possibility that millet crop grains were being obtained through trade and exchange. Ethnographic observations of plant processing particularly designed for archaeobotanical interpretations had been limited to Southwest Asian grasses (Hillman 1981, 1984; Jones 1987), and have never been done in a South Asian context or for millet crops. Two main ethnographic studies of millet crop processing in India were conducted: summer cultivation in Gujarat, northwest India, and 'Opportunistic' cultivation along a river floodplain in Andhra Pradesh, south India. The main objective of my ethnographic study was to discern the millet crop processing sequences, isolate variables that distinguish different stages, and develop models for identification of crop cultivation in archaeological contexts. An additional objective was to identify whether millets are processed differently depending on whether they are grown only as human food, both as human food and animal fodder, or when grown only as fodder.

The results of the ethnographic millet crop processing study are twofold: first, distinct variables can be isolated and used to distinguish the various processing stages; and second, there is a significant difference in processing pathways between different millets crops (Type A crops, *Sorghum bicolor* and *Pennisetum typhoides* versus Type B crops, *Panicum miliare* and *Setaria tomentosa*). The patterned variation between crop types and crop processing stages includes distinct compositions within the byproducts and products that are quantifiable. The differences in processing are a result of several factors including type of cultivation, crop type, ultimate use of product, seed morphology, and regional stylistic variation. There are a variety of harvesting choices depending on whether the crop is a Type A or a Type B variety (see chapter 4 or glossary). The choice of harvesting also affects the processing pathways with respect to the inclusion of weeds, the amount of byproduct cleaning, and the number of cleanings done at each stage. A major contribution of this study is the demonstration that harvesting styles (particularly the difference between Type I and Type II methods) have a direct affect on the compositions of the successive products and byproducts down the processing pathway.

The isolation of variables that distinguish the byproducts and products of specific millet crop processing stages, and the development of millet crop processing models (for Type A and Type B millet crops) specifically tailored for archaeological interpretations were the main themes of chapter 4. Ethnographic models were presented for the processing of Type A millet crops such as *Sorghum bicolor* and *Pennisetum typhoides,* and Type B crops such as *Setaria tomentosa* (growing with weeds) and *Panicum miliare* (growing without weeds). The models are flexible and facilitate any adjustments needed as a result of varying harvesting methods. Most importantly, the ethnographic models aid in the archaeological elucidation of cultivation versus procurement of grains from elsewhere.

The ethnographic crop processing models were rigorously tested on archaeobotanical assemblages from Babar Kot (very late Mature Harappan) and Oriyo Timbo (Late Harappan), through detailed study of archaeobotanical samples from varied contexts, and the role of millet cultivation was elucidated through the identification of crop cultivation and the economic importance of millets. In addition, defining the role of millets and their cultivation as animal fodder at Oriyo Timbo and Babar Kot aided in clarifying the interplay and integration between pastoral and agricultural economies during the very late third and second millennium B.C. in Gujarat.

The study revealed that the very late third millennium B.C. occupants at Babar Kot were engaged in year long local cultivation of summer and winter crops, including millets, legumes, and oilseed. Millets were the primary food grains and particular focus was on *Setaria italica*. They used millet grain as human food, and it appears to have been the primary food crop for the settlement and an integral component of the economy. In contrast, the Late Harappan Oriyo Timbo inhabitants were utilizing millets but not cultivating the crops at the site. Instead, they were either bringing grain in at the onset of occupation each season, or, trading for grain with neighboring agriculturalists. With respect to the issue of using millets as fodder, it appears that millets were brought to Oriyo Timbo as highly processed grain and hence it was unlikely to be used as fodder. The use of millets as fodder could not be definitively discerned at Babar Kot on the basis of the present data. However, there is a strong likelihood that millet byproducts were used as fodder by the occupants given the amount of byproduct generated during local millet cultivation, and the presence of domesticated animals within a farming context.

At Babar Kot, the practice of sedentary agriculture was supplemented by sedentary animal husbandry, as evidenced by the significantly high recovery of domesticated animal remains such as cattle, goat, sheep, and pig. Given that millet cultivation was a prominent activity and that domesticated animals were a significant commodity, the use of millet byproducts as animal fodder is highly probable, although wild fodder continued to be an important supplement. Additionally, Babar Kot was primarily a non-specialized settlement much like Rojdi C (and unlike specialized Harappan sites such as Kuntasi in North Gujarat, and Nageshwar in Saurashtra with their heavy reliance on trade) and the cultivation of cereals for consumption would have been crucial, since they would not be able to trade/exchange for food grain.

Given these results, several conclusions can be drawn regarding Babar Kot and its implications for late Mature Harappan sedentary communities in general. Cultivation of millets was an integral part of the subsistence economy at Babar Kot, and one can predict that sites of comparable features and in analogous settings in the area may have had a similar economy. However, the subsistence economy was not concentrated exclusively on cultivation, since wild plant and animal exploitation were significant components of the subsistence strategies, as was animal husbandry. The broad scope of the subsistence economy (ranging from farming, animal husbandry to wild plant and animal exploitation) is an important feature of Babar Kot. Of interest is the range of millets utilized at the site, which could be indicative of their differential use. Babar Kot millet cultivars included *Setaria italica, Panicum miliare* and *Eleucine coracana*, of which the last one had limited use as human food. The latter was probably only used as animal fodder, and cultivated specifically as green fodder.

The specialized use of particular crops is of great significance, and suggests a certain degree of complexity in the subsistence practices in terms of long term planning and specialized cultivation. For example, there is a great range in the millets recovered at the site of Rojdi, and of equal importance is the changing emphasis on different millets over time (Weber 1989). *Eleucine coracana* and *Panicum miliare* occur prominently in Rojdi A and B but decline in Rojdi C, while *Setaria* sp. and *Trianthema* sp. are dominant in Rojdi C but occur in very small numbers in Rojdi A and B (Weber 1992). This temporal change could be reflective of changing demands in the community, such as an increasing importance of millet byproducts or green millet crop for fodder. Such demands would necessitate the selection for millets, which have a higher percent of vegetative parts (to get higher mass of vegetative green fodder or byproduct). So, at Babar Kot there might have been a greater emphasis on *Eleucine coracana*, as compared to the other two millets, as green fodder for animals. Therefore, at

Babar Kot, the cultivation of millets as animal fodder (even minimally) and human food is suggestive of sedentary agricultural economy, with a very minor pastoral component, that of sedentary animal husbandry. It is conceivable that a separate group (an ethnic community distinctive from the farming segment) within the entire Babar Kot community was the only portion of the population practicing pastoralism. However, there is no evidence currently, archaeological or paleoethnobotanical, to support such a possibility.

At Oriyo Timbo, local cultivation of millets by the occupants was not identified. Instead, the millet grain was brought into the site as highly processed grain obtained either through trade or exchange with farming groups in the area, or cultivated elsewhere (outside the range of the settlement) by the occupants and brought along with them to Oriyo Timbo on their seasonal migrations. Since the Oriyo Timbo archaeobotanical assemblage does not include any millet byproduct assemblages, their use as fodder is improbable. Fodder for domestic animals would have comprised of wild forage selectively collected by the herders, or alternatively the animals could have been grazed and browsed freely. The human plant diet at Oriyo Timbo therefore included the use of millets and legumes, but was heavily supplemented by the collection of wild plants. Animal consumption (both wild and domesticated) also was an integral part of the subsistence strategies.

Given these conclusions, several broad implications can be drawn from Oriyo Timbo with respect to Late Harappan seasonal communities with pastoral emphasis. The subsistence economy of Oriyo Timbo can best be described as a form of semi-nomadic or semi-sedentary pastoralism, especially because it is possible that the occupants may have practiced millet cultivation elsewhere during other seasons. In the case of a semi-nomadic pastoralist economy, Oriyo Timbo inhabitants would have been primarily focused on animal husbandry and pastoral economy, but also conducted small-scale agriculture. Alternatively, in the case of a semi-sedentary pastoral economy, the occupants primarily practiced agriculture elsewhere (outside the region) but were also involved in pastoral activities seasonally at Oriyo Timbo.

It is important to stress that the subsistence economy at Oriyo Timbo was not exclusively focused on animal husbandry, since wild plant and animal exploitation were a significant component of their subsistence strategies. Therefore, in spite of the specialized economy, their subsistence base was relatively diverse. It is suggested that sites similar to Oriyo Timbo during the Late Harappan might have also shared a similar semi-nomadic or semi-sedentary pastoralist economy. It is important to reiterate that it is crucial to elucidate the integration between agriculture and pastoral economies.

The animal feeding model developed through ethnographic observations and archaeological inferences proposed that the diets of the Oriyo Timbo domesticated animals primarily consisted of wild forage, while the Babar Kot domesticated animals were fed millet byproducts when available (November onwards), and that wild forage supplemented their diet during other seasons. The animal feeding model can be tested when isotopic data from animal bones with sufficient collagen preservation are available, and at that point the model's implications of various isotopic signatures can be re-evaluated. The exact proportion of millets in animal diets is not as critical as relative long-term changes in the dependence on different fodders, particularly millets versus previously utilized wild fodders. An increased emphasis on millets as fodder during the Harappan Phase in Gujarat may indicate the emergence of a specialized subsistence system, with animal husbandry as an important and critical variable (Reddy 1991a, 1994). Such a change in subsistence economics, entailing specialized food production and intensification, could have far reaching implications for understanding the Gujarat Harappan Phase.

On a broader regional level, the results from the ethnoarchaeological and paleoethnobotanical study at Oriyo Timbo and Babar Kot have important implications for the interpretation of the Harappan subsistence systems and settlement patterns. Bhan (1989) and Possehl (1992) have suggested that the Harappan in Gujarat typically shows a development of two different categories of settlements: small settlements most often interpreted as small villages or temporary encampments of migratory herding populations; and larger sedentary settlements with more permanent architecture. Artifactual ties between these two categories of settlements are absent at present, however this two-tier system has interpretive implications particularly if there are functional differences - the smaller possible seasonal settlements being pastoral and the larger sedentary settlements are agricultural.

The settlement and subsistence system would be differentiated on the basis of the degree of reliance on agriculture by pastoralists and the relationship between pastoralists and agriculturalists. To define such systems archaeologically, it is imperative that cultivation and processing of crops (if any), including their growing seasons (as an economic activity) are elucidated at a given settlement. However, there are several other parameters besides cultivation that are indicative of the different pastoral adaptations, since cultivation is practiced to varying degrees depending on the type of pastoral adaptation. These include seasons of occupation, site size, permanence and style of architecture, layout and organization of the settlement (especially intra-site patterning to identify ethnically distinct groups), and artifact density. Additionally, the diversity of site types within the settlement system will be a reflection of different adaptations. For example, in a system where there are limited variations in adaptations (involving only sedentary agriculturalists and herdsman husbandry pastoralists) there is little settlement diversity consisting only of permanent agricultural settlements and uniform pastoral camps. Typically settlements of herdsman husbandry pastoralists have minimal architectural preservation and low diversity in artifact categories, while the sedentary agricultural settlements would have high architectural visibility including substantial structures and a wider range of artifact categories. In contrast, a settlement system with a more diverse range of groups including either semi-sedentary or semi-nomadic pastoralists will have a greater diversity of site types. These sites will have different seasonal indicators with respect to the nature of the architecture, site size, the degree of animal culling, and so on.

These ethnoarchaeological studies of traditional crop processing conducted in Gujarat and Andhra Pradesh have provided new crop processing models with global applicability and new avenues for addressing and elucidating crop cultivation, fodder utilization, and food processing for the Harappan Phase. Research into the role of agricultural products in highly developed economies enriches our understanding of complex societies, particularly with respect to the degree of integration present between aspects of the population. When this approach is applied to additional Harappan Phase sites in Gujarat, further insights into the complex interrelationship between pastoralism and agriculture should be forthcoming.

Glossary

This glossary provides information for key words and concepts used in the text, but it is not a comprehensive dictionary of archaeology or botany. An effort was made in the text to keep jargon to a minimum, although a few technical expressions are inevitable.

Caryopsis: the botanical term for a dry one-seeded fruit with the pericarp and testa fused together. Other synonymous terms used for caryopsis in this book include seed and grain.

Chaff: the crop component produced when the glumes/husks of the grains are removed through pounding or grinding. It is dusty in composition and preserves poorly in archaeological contexts.

Cultigen: a plant species or other taxon evolved under domestication to become distinct from its wild progenitor or progenitors.

Cultivar: a particular variety of a cultigen, usually named informally. Varieties or strains of cultivated crops, not forming botanical 'varieties'.

Free threshing: a term used in relationship to some cereals to describe the ease of dislodging the grain/caryopsis from the spikelets and glumes. The grains of such cereals are freed from their enclosing spikelets only through threshing. Examples of free threshing millets studied in this project are *Sorghum bicolor* and *Pennisetum typhoides*.

Heads/ Inflorescence heads/ Panicles/ Panicle heads: all these terms are synonymous in thsi book, and refer to the seed bearing part of the grass plant. In the cases of millet crops, the seed bearing portion is a specific inflorescence form called a panicle. The panicle heads bear the seed bearing spikelets on rachilla attached on the central rachis.

Headedness: the tendency for the grains (of both crop plants and weeds) to remain on the head/ seed bearing structure (the panicle, or rachis, or rachilla) in spite of threshing.

Processing stages levels: Relative terminology is used to distinguish levels of processing stages:

Higher processing stages are those stages which are closer to the consumption product or end product such as pounding, winnowing II and grinding. They also most often occur primarily in home bases or domestic activity areas.

Lower processing stages refer to the initial stages of processing such as threshing, winnowing I and sieving, occur in the fields and home bases.

Palea and Lemma/Husk/Glumes: the outer coverings of the caryopsis/seeds/grains of grasses. These coverings may be tightly bound and hard to remove (as seen for *Setaria italica* and *Panicum miliare*) or soft and easy to remove (as seen for *Sorghum bicolor* and *Pennisetum typhoides*).

Rachis(ii): the portion of the central stalk of the grass plant that lies within the panicle. The spikelets which bear the grains are attached to the rachis either directly or through rachillas.

Rachilla(s): secondary axes on the rachis and terminate in numerous spikelets which bear the seeds.

Sievability: the index of how easily a seed can be ethnographically sieved/separated away from of the prime products. It is dependent on seed size and weight.

Spikelets: consist of the crop grains enclosed within the glumes/husks and often held on the ends of rachii or rachillas. The spikelets have to be threshed/pounded to free the grains. In the case of *Sorghum bicolor* and *Pennisetum typhoides* when the grains fall free of the rest of the spikelet they are most often naked (i.e., with no enclosing glume). In other cases, they have glumes but not the rachilla portion of the spikelet. For *Panicum miliare* and *Setaria tomentosa*, when the grain falls free it often has part of the rachilla attached, and only pounding totally frees the grain.

Type A crops: the term given to crops that can be harvested in a particular manner, termed Type I method of harvest. Examples include *Pennisetum typhoides*, *Sorghum bicolor*. The processing of these crops is similar to that for free threshing cereals (as discussed by Hillman 1983).

Type I harvest method: typically involves only cutting the panicle heads, leaving the stalks behind. The harvest cut occurs at the base of the panicle heads, and this represents a specific selection for only the seed bearing parts of the plants as the starting product of processing.

Type B crops: include small grained plant crops such as *Setaria italica* and *Panicum miliare* which are only harvested by a particular method, termed Type B method of harvest. The processing of these crops is similar to that for glume wheats (as discussed by Hillman 1983).

Type II harvest method: involved the cutting of the panicle heads, the stalks, and most often they are cut at the base of the plant. This method incorporates weeds growing in between and around the crop plants into the processing stages. The starting product for processing includes seed bearing parts and non-seed bearing parts also. It is important to note that Type A crops can be harvested in a Type II method.

Winnowability: the index of how easily a seed is separated from the prime products through winnowing. It is dependent on seed size and weight.

References

Agrawal, D. P., and B. M. Pande (eds)
1977 *Ecology and Archaeology of Western India*. Concept Publishing Co., Delhi.

Agrawal, D. P., and R. K. Sood
1982 Ecological Factors and the Harappan Civilization. In *Harappan Civilization*, edited by G. L. Possehl, pp. 223-52. American Institute of Indian Studies and Oxford & IBH Pub. Co., Delhi.

Aiyer, A. K. Y. N.
1982 *Field Crops of India*. Bangalore Printing and Publishing Company Ltd., Bangalore.

Allchin, B., and A. Goudie
1971 Dunes, Aridity and Early Man in Gujarat, Western India. *Man* 6:248-65.

Allchin, B., A. Goudie, and K. Hegde
1978 *The Prehistory and Paleogeography of the Great Indian Desert*. Academic Press, New York.

Barnard, G., and Kristoferson, L.
1989 *Agricultural Residues as Fuel in the Third World*. Earthscan, Energy Information Programme Technical Report No. 4. The Beijer Institute, The Royal Swedish Academy of Science.

Bhan, Kuldeep K.
1989 Late Harappan Settlements of Western India with Specific Reference to Gujarat. In *Old Problems and New Perspectives in the Archaeology of South Asia*, edited by J. M. Kenoyer, pp. 61-74, Wisconsin Archaeological Reports, Vol. 2.
1992 Late Harappan Gujarat. *The Eastern Anthropologist* 45(1-2):173-192.

Bhandari, M.N.
1974 Famine Foods in the Rajasthan Desert. *Economic Botany* 28:73-81.
1978 *Flora of the Indian Desert*. Scientific Publishers, Jodhpur.

Bharadwaj, O.P.
1961 The Arid Zone of India and Pakistan. In *A history of Land use in Arid Regions*, edited by L.D. Stamp, pp. 143-74. UNESCO, Arid Zone Research, Paris.

Binford, L. R.
1981 *Bones: Ancient Man and Modern Myths*. Academic Press, New York.

Bisht, R. S.
1990 Dholavira: New Horizons of the Indus Civilization. *Puratattva* 20:71-82.

Bottema, S.
1984 The Composition of Modern Charred Seed Assemblages. In *Plants and Ancient Man*, edited by W. van Zeist and W. A. Casperie, pp. 207-212. A. A. Balkema, Rotterdam.

Cane, S.
1989 Australian Aboriginal Seed Grinding and its Archaeological Record: A Case Study from the Western Desert. In *Foraging and Farming: The Evolution of Plant Exploitation*, edited by D. R. Harris and G. C. Hillman, pp. 99-119. Unwin Hyman, London.

Caswell, H., F. Reed, S. N. Stephenson, and P. A. Werner.
1973 Photosynthetic Pathways and Selective Herbivory: A Hypothesis. *The American Naturalist* 107 (956):465-480.

Chakrabarti, D. K.
1984 Origins of the Indus Civilization: Theories and Problems. In *Frontiers of the Indus Civilization*, edited by B. B. Lal and S. P. Gupta, pp. 43-40. Indian Archaeological Society, New Delhi.

Chalam, G.V., and J. Venkateswarlu
1965 *An Introduction to Agricultural Botany in India*. Vol I. ASIA, Bombay.

Charles, M. P.
1990 Traditional Crop Husbandry in Southern Iraq. *Bulletin on Sumerian Agriculture* 5:47-64.

Chisholm, B. S., D. E. Nelson and H. P. Schwartz
1983 Marine and Terrestrial Protein in Prehistoric Diets on the British Columbia Coast. *Current Anthropology* 24:396-398.

Cobley, L. S.
1956 *An Introduction to the Botany of Tropical Crops*. Longmans Green and Co., London.

Dales, G. F.
1979 The Balakot Project: Summary of Four Years' Excavations in Pakistan. In *South Asian Archaeology 1977*, edited by M. Taddei, pp. 241-73. IsMEO, Naples.

Dales, G. F. and J. M. Kenoyer.
1990 Excavation at Harappan - 1988. *Pakistan Archaeology* 24:68-176.
1991 The Harappa Project 1986-1989: New Investigations at an Ancient Indus City. In *Harappan Civilization: A Contemporary Perspective*, edited by G. L. Possehl. 2nd edition. Oxford and IBH, New Delhi.

DeNiro, M.
1985 Postmortem Preservation and Alteration of in vivo Bone Collagen Isotope Ratios in Relation to Paleodietary Reconstruction. *Nature* 317:806-809.

DeNiro, M., and S. Epstein.
1978 Influence of Diet on Distribution of Carbon Isotopes in Animals. *Geochemica et Cosmochemica Acta* 42:495-506.
1981 Influence of Diet on the Distribution of Nitrogen Isotopes in Animals. *Geochemica Cosmochemica Acta* 45:341-351.

Dennell, R. W.
1972 The Interpretation of Plant Remains: Bulgaria. In *Papers in Economic Prehistory*, edited by E. S. Higgs, pp.149-50. Cambridge University Press, Cambridge.
1974 Prehistoric Crop Processing Activities. *Journal of Archaeological Science* 1:275-84.
1976 The Economic Importance of Plant Resources Represented on Archaeological Sites. *Journal of Archaeological Sciences* 3:229-247.

Dhavalikar,
1993 New Perspectives in Harappan Studies. *Man and Environment* XVIII (2):87-91.

Dikshit, R. B. K. N.
1938 *Prehistoric Civilization of the Indus Valley*. University of Madras Press, Madras.

Dixit, A.S., S.S. Dixit and Vishnu-Mittre.
1987 The Occurrence of *Eleucine africana* Kennedy O'Byrne in India and Its Significance in the Origin of *Eleucine coracana*. *Proceedings of the Indian Academy of Sciences (Plant Sciences)* 97(1):1-10.

Dodia, R.
1988 Climate of Kashmir during the last 700,000 years: The Baltal Pollen Profile. *Proceedings Indian National Science Academy* 54 A(3):481-89.

Dodia, R., A. C. Gupta, C. Mandavia, C. Charma and A. Vora
1982 Palynological Investigations on the Lower Karewas, Kashmir. *Man and Environment* 6:21-26.

Fairservis, W. A.
1956 Excavations in the Quetta Valley, West Pakistan. *Anthropology Papers of the American Museum of Natural History* 45 (part 2).
1973 Preliminary Report on Excavations at Alladino (First Season, 1973). *Pakistan Archaeology*. 9:95-102.

Flam, L.
1981 *The Paleography and Prehistoric Settlement Patterns in Sind, Pakistan (ca 4000-2000 B.C.)*, PhD Dissertation, University of Pennsylvania, Philadelphia.

Gosh, S. S., and K. R. Lal.
1963 Plant Remains from Rangpur. *Ancient India* 18-19:161-77.

Government of India.
1908 *Imperial Gazetteer of India*. Vol 15. (New Edition).

Gupta, H.P.
1971a Pollen Analytical Investigations of Some Upper Pleistocene Samples from Tocklai, Cinnamara, Assam. *The Paleobotanist* 18(3):234-236.
1971b Studies of Indian Pollen Grains -IV Boraginaceae. *Geophytology*: pp.127-34.

Gupta, S.K.
1977 Holocene setting in the Little Rann of Kutch. In *Ecology and Archaeology of Western India*, edited by O.P Agrawal and B.M. Pande, pp. 201-205. Concept Publishing Company, Delhi.

Gupta, H.P., and C. Sharma
1982 Quaternary Palynostratigraphy in India: A Critical review. *Special Publication of the Palaeontological Society of India*. 1:130-38.

Gupta, S. P.
1979 *Archaeology of Soviet Central Asia and the India Borderlands*. 2 vols. B.R. Publishing Co., Delhi.

Hastorf. C. A., and V. S. Popper (eds).
1988 *Current Paleoethnobotany: Analytical Methods and Cultural Interpretations of Archaeological Plant Remains*. The University of Chicago Press, Chicago and London.

Heath, M. E., R. F. Barnes and D. S. Metclalfe.
1985 *Forages: The Science of grassland agri-*

culture, 4th edition. Iowa State University Press, Ames.

Hegde, K.

1977 Late Quaternary Environments in Gujarat and Rajasthan. In *Ecology and Archaeology of Western India*, edited by D. P. Agrawal and B. M. Pande, pp. 169-80. Concept Publishing Co., Delhi.

Hegde, K. T. M., V. H. Sonawane, D. R. Shah, K.K. Bhan, Ajitprasa, K. Krishnan and A. Pratapa Chandran

1988 Excavation at Nagwada - 1986 and 1987: A Preliminary Report. *Man and Environment* 12:55-65.

Herman, Frank

1991 *Bronze Age Complexes in Gujarat Before the Harappan Urbanization Expansion: A Working Hypothesis.* Master Thesis, Department of Anthropology, University of Pennsylvania, Philadelphia.

1997a "Harappan" Gujarat: The Archaeology-Chronology Connection. *Paleorient,* 22(2):77-112.

1997b The Rangpur Sequence of "Harappan" Gujarat (India): A Re-Assessment. In *South Asian Archaeology, 1995,* edited by Raymond Allchin and Bridget Allchin, pp. 187-198. Oxford & IBH Publishing Co., Pvt. Ltd., New Delhi.

Hillman, G.

1981 Reconstructing Crop Husbandry Practices from the Charred Remains of Crops. In *Farming Practice in British Prehistory*, edited by R. J. Mercer, pp.123-162. Edinburgh University Press, Edinburgh.

1984 Interpretation of Archaeological Plant Remains: the Application of Ethnographic Models from Turkey. In *Plants and Ancient Man: Studies in Paleoethnobotany*, edited by W.V. Zeist and W.A. Casperie, pp.1-41. A. A. Balkema, Rotterdam.

1985 Traditional Husbandry and Processing of Archaic Cereals in Recent Times: the Operations, Products and Equipments That Might Feature in Sumerian Texts. Part I: The Glume Wheats. *Bulletin on Sumerian Agriculture* 1:114-152. Part II: The Free Threshing Cereals. *Bulletin on Sumerian Agriculture* 2:1-31.

Hillman, G. C., and M. S. Davies.

1990 Measured Domestication Rates in Wild Wheats and Barley Under primitive Cultivation, and their Archaeological Implications. *Journal of World Prehistory*, vol. 4, No.2:157-222.

Ho, P.

1975 *The Cradle of East.* University of Chicago Press, Chicago.

Hodder, I.

1982 *The Present Past.* B. T. Batsford, London.

Hutchinson, J

1977 India: Local and Introduced Crops. In *The Early History of Agriculture*, edited by J. Hutchinson, G. Clarke, E. M. Jope and R. Riley. pp. 129-141. Oxford University Press for the British Academy, Oxford.

Jacobson, J.

1987 The Harappan Civilization: An Early State. In *Studies in the Archaeology of India and Pakistan,* edited by J. Jacobson, pp. 137-174. Oxford and IBH, New Delhi.

Jarrige, J. F.

1986 Excavation at Mehrgarh-Nausharo. *Pakistan Archaeology* 10-22:62-131.

1988 Excavations at Nausharo. *Pakistan Archaeology* 23:149-203.

1997 From Nausharo to Pirak: Continuity and Change in the Kachi/Bolan Region from the 3rd to the 2nd Millennium B.C. In *South Asian Archaeology, 1995,* edited by Raymond Allchin and Bridget Allchin, pp. 11-32. Oxford & IBH Publishing Co., Pvt. Ltd., New Delhi.

Jarrige, J. F., and R. H. Meadow.

1980 The Antecedents of Civilization in the Indus Valley. *Scientific American* 243 (2):122-133.

Jones, G.

1983 The Use of Ethnographic and Ecological Models in the Interpretation of Archaeological Plant Remains: Case Studies from Greece. Unpublished Dissertation for the University of Cambridge.

1984 Interpretation of Archaeological Plant Remains: Ethnographic Models from Greece. In *Plants and Ancient Man: Studies in Paleoethnobotany*, edited by W.V. Zeist and W.A. Casperie, pp. 43-61. A. A. Balkema, Rotterdam.

1987 A Statistical Approach to the Archaeological Identification of Crop Processing. *Journal of Archaeological Science* 14:311-323.

1988 The Application of Present Day Cereal Processing Studies to Charred Archaeobotanical Remains. *Circeae* 6 (2):91-96.

Jones, G. E. M., K. A. Wardle, P. L. J. Halstead and D. Wardle.
1986 Crop Storage at Assiros. *Scientific American* 254 (3):96-103.
Joshi, J. P., M. Bala, and J. Ram.
1984 The Indus Civilization: A Reconsideration on the Basis of Distribution Maps. In *Frontiers of the Indus Civilization*, edited by B. B. Lal and S. P. Gupta, pp. 511-530. Books and Books, New Delhi.
Khazanov, A.
1984 *Nomads and the Outside World*. Cambridge University Press, Cambridge.
Kenoyer, J. M.
1989 Socio-economic Structure of the Indus Civilization as Reflected in Specialized Crafts and the Question of Ritual Segregation. In *Old Problems and New Perspectives in the Archaeology of South Asia*, edited by J.M. Kenoyer, pp.183-192. Wisconsin Archaeological Reports, Vol. 2.
1991 The Indus Valley Tradition of Pakistan and Western India. *Journal of World Prehistory* Vol.5, 4:331-385.
1992 Harappan Craft Specialization and The Question of Urban Segregation and Stratification. *The Eastern Anthropologist* 45(1-2):39-54.
1997a Trade and technology of the Indus Valley: Insights from Harappa, Pakistan. *World Archaeology*, 29(2):292-280.
1997b Early City-States in South Asia: Comparing the Harappan Phase and Early Historic Period. In *The Archaeology of City-States, Cross Cultural Approaches*, edited by Deborah L. Nichols and Thomas H. Charleston, pp.51-70. Smithsonian Institution Press, Washington D.C.
Kramer, C.
1979 *Ethnoarchaeology: Implications of Ethnography for Archaeology*. Columbia University Press, New York.
Lamberg-Karlovsky, C. C.
1982 Sumer, Elam and the Indus: Three Urban Processes Equal One Structure? In *Harappan Civilization: A Contemporary Perspective*, edited by G.L.Possehl.pp. 61-68. Oxford and IBH Pub. Co., New Delhi.
Leshnik, L.S
1968 Prehistoric Exploration in North Gujarat and Parts of Rajasthan. *East and West* 18 (3-4):295-310.
Mackay, E. J. H.
1943 *Chanhu-daro Excavation 1935-36*. Boston Museum of Fine Arts, American Oriental Series.
Mann, H.H.
1955 *Rainfall and Famine: A study of rainfall in the Bombay Deccan, 1865-1938*. Indian Society of Agricultural Economies, Bombay.
Marshall, S. J.
1931 *Mohenjo-daro and the Indus Civilization*. A. Probsthain, London.
Meadow, Richard
1989 Continuity and Change in the Agriculture of the Greater Indus Valley: The Paleobotanical and Zooarchaeological Evidence. In *Old Problems and New Perspectives in the Archaeology of South Asia*, edited by J.M. Kenoyer, pp.61-74. Wisconsin Archaeological Reports, Vol. 2.
1996 The Origins and spread of agriculture and pastoralism in northwestern South Asia. In *The Origins and Spread of Agriculture and Pastoralism* in Eurasia, edited by David Harris, pp. 390-412. Smithsonian Institution Press, Washington D.C.
Meadow, Richard and Mark Kenoyer
1997 Excavations at Harappa 1994-1995: New Perspectives on the Indus Script, Craft Activities, and City Organization. In pp. In *South Asian Archaeology, 1995*, edited by Raymond Allchin and Bridget Allchin, pp. 139-172 . Oxford & IBH Publishing Co., Pvt. Ltd., New Delhi.
Meher-Homji, V.M.
1989 History of Vegetation of Peninsular India. *Man and Environment* 13:1-10.
Mehra, K. L.
1963 Considerations of the African Origin of *Eleucine* coracana (L.) Gaertn. *Current Science* 7:300-301.
Miksicek, C.
1983 Archaeobotanical Aspects of Las Flosas: A Statistical Approach to Prehistoric Plant Remains. In *Hohokam Archaeology Along the Salt-Gila Aqueduct*, Vol. I, parts V and VI, edited by L.S. Teague and P. L. Crown, pp. 671-700. Arizona State Museum Archaeology Series No. 150.
1987 Formation Processes of the Archaeobotanical Record. *Advances in Archaeological Method and Theory* 10:211-247. Academic Press, San Diego.
Miller, N. F.
1984 The Use of Dung as Fuel: An Ethno-

graphic Example and An Archaeological Application. *Paleorient* vol. 10/2:71-79.
1988 Ratios in Paleoethnobotanical Analysis. In *Current Paleoethnobotany: Analytical Methods and Cultural Interpretations of Archaeological Plant Remains*, edited by Christine A. Hastorf and Virginia S. Popper, pp. 72-85. University of Chicago Press, Chicago.

Miller, N. F., and T. L. Smart.
1984 Intentional Burning of Dung Fuel: A Mechanism for the Incorporation of Charred Seeds into the Archaeological Record. *Journal of Ethnobiology* 4:15-28.

Minnis, Paul
1978 Paleoethnobotanical Indicators of Prehistoric Disturbance: A Case Study. In *The Nature and Status of Ethnobotany,* edited by Richard Ford, M. Brown, M. Hodge, and W. L. Merril, pp. 347-66. Anthropological Papers no. 67. University of Michigan, Museum of Anthropology, Ann Arbor.

Mughal, M. R.
1982 Recent Archaeological Evidence from Bhawalpur. *Man and Environment* 4:93-98.
1990 The Decline of the Indus Civilization and the Late Harappan Period in the Indus Valley. *Lahore Museum Bulletin* 111(2): 1-17.

O'Leary, Marion
1981 Carbon Isotope Fractionation in Plants. *Phytochemistry* 20(4):553-367.

Patel, Ajita
1997 The Pastoral Economy of Dholavira: A First Look at Animals and Urban Life in Third Millennium Kutch. In *South Asian Archaeology, 1995*, edited by Raymond Allchin and Bridget Allchin, pp. 101-114. Oxford & IBH Publishing Co., Pvt. Ltd., New Delhi.

Patel, G.
1977 *Gujarat's Agriculture.* Overseas Book Traders, Ahmedabad.

Pearsall, D. M.
1988 Interpreting the Meaning of Macroremain Abundance: The Impact of Source and Context. In *Current Paleoethnobotany: Analytical Methods and Cultural Interpretations of Archaeological Plant Remains*, edited by Hastorf. C. A., and V. S. Popper. pp. 97-118. University of Chicago Press, Chicago and London.
1982 Phytolith Analysis: Applications of a New Paleoethnobotanical Technique in Archaeology, *American Anthropologist* 84:862-871.

Piggott, S.
1962 *Prehistoric India* (2nd edition). Cassell, London.

Possehl, Gregory L.
1976 Lothal: A Gateway Settlement of the Harappan Civilization. In *Ecological Backgrounds of South Asian Prehistory*, edited by K.A.R. Kennedy and G.L. Possehl, pp. 118-31. Ithaca: Cornell University, South Asia Occasional Papers and Theses, No. 4.
1979 Pastoral Nomadism in the Indus Civilization: An hypothesis. In *South Asian Archaeology, 1977*, edited by Taddei, M., pp. 537-551. Istituto Universitario Orientale, Naples.
1980 *Indus Civilization in Saurashtra*. B. R. Publishing Corp., Delhi.
1986 African Millets in South Asian Prehistory, In *Studies in the Archaeology of India and Pakistan*, edited by Jerome Jacobson, pp.137-174. Oxford & IBH, New Delhi.
1990 *A Report on the Excavations at Oriyo Timbo.* Report submitted to the Government of India.
1992 The Harappan Civilization in Gujarat: The Sorath and Sindhi Harappans. *The Eastern Anthropologist* 45(1-2):117-154.
1997 The transformation of the Indus Civilization. *Journal of World Prehistory* 11(4):425-472.

Possehl, G.L. and F. Herman
1990 The Sorath Harappan: A new regional manifestation of the Indus urban phase. In *South Asian Archaeology 1987*, edited by M. Taddei and P. Calliero, pp. 295-319. Istituto Italiano per il Medio ed Estremo Oriente, Serie Orientale 66(1), Rome.

Possehl, G.L. and D. P. Mehta
1994 Excavation at Rojdi, 1992-1993. In *South Asian Archaeology 1993*, edited by Parpola A. and P. Koskikallio, pp. 603-614. Annales Academiae Scientiarum Fennicae, Series B., Vol. 271, Helsinki.

Possehl, G. L., and M. H. Raval.
1989 *Harappan Civilization and Rojdi.* American Institute of Indian Studies. and Oxford & IBH Publishing Company Pvt. Ltd., New Delhi.

1991 *Excavations at Babar Kot 1991*. Report submitted to the Government of India.

Purseglove,
1979 *Tropical Crops: Monocotyledons.* Longman, London.

Rachie, K., and J. V. Majmudar.
1980 *Pearl Millet.* The Pennsylvania State University Press, University Park & London.

Raikes, R. L., and R. H. Dyson.
1961 The Prehistoric Climate of Baluchistan and the Indus Valley. *American Anthropologist* 63:265-281.

Rao, K. E., J. M. J. de Wet, B. E. Brink and M. H. Mengesha
1987 Intraspecific Variation and Systematics of Cultivated *Setaria italica*, Foxtail millet, *Economic Botany* 41(1):108-116.

Rao, S.R.
1973 *Lothal and the Indus Civilization.* Asia Publishing House, Bombay.

Rathje, W. L., and M. B. Shiffer.
1982 *Archaeology.* Harcourt, Brace, New York.

Reddy, Seetha N.
1991a Complementary Approaches to Late Harappan Subsistence: An Example from Oriyo Timbo. In *The Archaeology of 3rd millennium Urbanism at Harappa (Pakistan)*, edited by Richard Meadow, pp. 127-135 Prehistory Press, Madison.
1991b Archaeobotany at Oriyo Timbo 1989-90: A post urban site in Gujarat. *Man and Environment* Vol.16:1, pp. 73-84.
1991c Opportunistic Cultivation: An Ethnographic Study in South India. *Expedition* Vol.33 (3):18-26.
1994 *Plant Usage and Subsistence Modeling: An Ethnoarchaeological Approach to the Late Harappan of Northwest India.* University Microfilms, Ann Arbor.
1997 If the Threshing Floor Could Talk: Integration of Agriculture and Pastoralism During the Late Harappan in Gujarat, India. *Journal of Anthropological Archaeology,* 16 (2):162-187.
1999 Fueling the Hearths in India: The Role of Dung in Paleoethnobotanical Interpretation. *Paléorient* 24 (2) 61-70.

Renfrew, J. M.
1973 *Paleoethnobotany.* Columbia University Press, New York.

Rissman, P.
1985 Migratory Pastoralism in Western India in the Second Millennium B.C.: The Evidence from Oriyo Timbo. PhD Dissertation, Department of Anthropology, University of Pennsylvania.
1985b The Oriyo Test Excavation and the End of the Harappan Tradition in Gujarat. *South Asia Archeology 1983,* edited by J. Schotsmans and M. Taddei, pp. 345-355. Universitario Orientale, Dipartmento di Studi Asiatici, Series Minor XXIII, Naples.
1987 The Potential for Annular Analysis for South Asian Archaeology. *Man and Environment* XI:15-24.

Rissman, P.C., and Y.M. Chitalwala.
1990 *Harappan Civilization and Oriyo Timbo.* American Institute of Indian Studies and Oxford &IBH Publishing Company Pvt. Ltd., New Delhi.

Ryan, K.
1990 The Faunal Remains from Oriyo Timbo, Gujarat. In *A Report on the Excavations at Oriyo Timbo*, edited by G. L. Possehl, Report submitted to the Government of India.
1991 Archaeozoology. In *Excavations at Babar Kot 1991*, edited by G. L. Possehl and M. H. Raval, Report submitted to the Government of India.

Sankalia, H. D.
1972 *Archaeological Excavations at Somnath, Guide Book.* Deccan College Post-Graduate and Research Institute, Poona.

Savitri, R., and Vishnu-Mittre.
1978 *Setaria* sps in Ancient plant economy of India. *The Palaeobotanist* 25:539-62.

Schiffer, Michael B.
1976 *Behavioral Archaeology.* Academic Press, New York.
1983 Toward the Identification of Formation Processes. *American Antiquity* 48 (4): 675-706.

Schoeninger, M. J., and M. J. DeNiro.
1984 Nitrogen and Carbon Isotope Composition of Bone Collagen from Marine and Terrestrial Animals. *Geochemica et Cosmochemica Acta* 48:625-637.

Schwarcz, H. P., and M. J. Schoeninger.
1991 Stable Isotope Analysis in Human Nutritional Ecology. *Yearbook of Physical Anthropology* 34:283-321.

Sealy, J. C., and N. J. Van der Merwe.
1986 Isotope Assessment and the Seasonal Mobility Hypothesis in the South Western Cape of South Africa. *Current Anthropology* 27(2):135-150.

Shaffer, J. G.
1981 The Protohistoric Period in the Eastern

Punjab: A Preliminary Assessment. In *Indus Civilization*, edited by A. H. Dani, pp. 65-102. Qaid-i-Azam University, Islamabad.

1991 The Indus Valley, Baluchistan and Helmand Traditions: Neolithic through Bronze Age. In *Chronologies in Old World Archaeology*, edited by R. Ehrich, 3rd edition, Vol. 1:441-464. University of Chicago Press, Chicago.

Shaffer, J. G., and D. A. Lichtenstein.

1989 Ethnicity and Change in the Indus Valley Cultural Tradition. In *Old Problems and New Perspectives in the Archaeology of South Asia*, edited by J.M. Kenoyer, pp. 117-126. Wisconsin Archaeological Reports, Vol. 2.

Sharma, A.K.

1992 The Harappan Horse was Buried Under the Dunes of....... *Puratattva* 23:30-34.

Singh, G.

1971 The Indus Valley Culture Seen in the Context of Post-Glacial Climatic and Ecological Studies in North-west India. *Archaeology and Physical Anthropology in Oceania* 6(2):177-89.

1977 Stratigraphic and Palynological Evidence for Desertification in the Great Indian Desert. *Annals of the arid Zone* 16 (3):310-20.

Spate, O.H.K., and A.T.A. Learmouth.

1967 *India and Pakistan. A General and Regional Geography*. (3rd edition). Methuen, London.

Stiles, D.

1977 Ethnoarchaeology: A Discussion of Methods and Applications. *Man*, 12(1):87-103.

Tuross, N., M. L. Fogel, and P.E. Hare.

1988 Variability in the Preservation of the Isotopic Composition of Collagen from Fossil Bone. *Geochemica et Cosmochemica Acta* 52:929-935.

Van Zeist, W., and W. A. Casperie (eds).

1984 *Plants and Ancient Man: Studies in Paleoethnobotany*. A. A. Balkema, Rotterdam.

Van der Merwe, N. J.

1982 Carbon Isotopes, Photosynthesis and Archaeology. *American Scientist* 70:596-606.

Vavilov, N. I.

1959 *The Origin, Variation and Breeding of Cultivated Plants*. Ronald Press Co., New York.

Vishnu-Mittre and R. Savitri

1982 Food Economy of the Harappan. In *Harappan Civilization*, edited by G. L. Possehl, pp. 205-222. American Institute of Indian Studies and Oxford & IBH, Delhi.

Wadia, D.N

1966 *Geology of India*. Third Edition. Macmillan, London.

Wagner, G.E

1982 *Flotation for Small-artifact Recovery at Oriyo Timbo*. University Museum, University of Pennsylvania, Philadelphia.

1985 Botanical Remains. In *Preliminary Report: Excavations at Rojdi 1983-84*, edited by G. L. Possehl. *Man and Environment* Vol IX.

Weber, S.

1992 South Asian Archaeological Variability. In *South Asian Archaeology 1989,* edited by C. Jarrige with assistance of John P. Gerry and Richard H. Meadow, pp. 283-290. Monographs in World Archaeology No. 14. Prehistory Press, Madison.

1991 *Plants and Harappan Subsistence*. American Institute of Indian Studies. Oxford & IBH Publishing Co. Pvt. Ltd., New Delhi.

1989 Plants and Harappan Subsistence: An example of Stability and Change from Rojdi. Ph.D. Dissertation for the Department of Anthropology, University of Pennsylvania, Philadelphia.

1997 Harappan Archaeobotany: A Model for Subsistence. In *South Asian Archaeology, 1995,* edited by Raymond Allchin and Bridget Allchin, pp. 115-118. Oxford & IBH Publishing Co., Pvt. Ltd., New Delhi.

1998 Out of Africa: The Initial Impact of Millets in South Asia. *Current Anthropology* 39 (2):267-274.

Werth E.

1937 Geography and History of Millets. *Angewandte Botanik* 19:41.

Wilhemy, H.

1969 Das Urstromtal am Ostrand der Iduse bene und das Sarasvati Problem. *Zeitschrift fur Geomorphologie* 8:76-91.

Wilson, J. R. and C. W. Ford.

1971 Temperature Influences on the Growth, Digestibility and Carbohydrate Composition of Two Tropical Grasses *Panicum maximum* var *trichoglume* and *Setaria spaceolata*, and Two Cultivars of the Temperate Grass *Lolium perenne*. *Australian Journal of Agricultural Research* 22:563-571.